Practical Web Scraping for Data Science

Best Practices and Examples with Python

Seppe vanden Broucke
Bart Baesens

Apress®

Practical Web Scraping for Data Science

Seppe vanden Broucke
Leuven, Belgium

Bart Baesens
Leuven, Belgium

ISBN-13 (pbk): 978-1-4842-3581-2
https://doi.org/10.1007/978-1-4842-3582-9

ISBN-13 (electronic): 978-1-4842-3582-9

Library of Congress Control Number: 2018940455

Managing Director, Apress Media LLC: Welmoed Spahr
Acquisitions Editor: Todd Green
Development Editor: James Markham
Coordinating Editor: Jill Balzano

Cover designed by eStudioCalamar

Cover image designed by Freepik (www.freepik.com)

Distributed to the book trade worldwide by Springer Science+Business Media New York, 233 Spring Street, 6th Floor, New York, NY 10013. Phone 1-800-SPRINGER, fax (201) 348-4505, e-mail orders-ny@springer-sbm.com, or visit www.springeronline.com. Apress Media, LLC is a California LLC and the sole member (owner) is Springer Science + Business Media Finance Inc (SSBM Finance Inc). SSBM Finance Inc is a **Delaware** corporation.

For information on translations, please e-mail rights@apress.com, or visit http://www.apress.com/rights-permissions.

Apress titles may be purchased in bulk for academic, corporate, or promotional use. eBook versions and licenses are also available for most titles. For more information, reference our Print and eBook Bulk Sales web page at http://www.apress.com/bulk-sales.

Any source code or other supplementary material referenced by the author in this book is available to readers on GitHub via the book's product page, located at www.apress.com/9781484235812. For more detailed information, please visit http://www.apress.com/source-code.

Printed on acid-free paper

Dedicated to our partners, kids and parents.

Table of Contents

About the Authors

Seppe vanden Broucke is an Assistant Professor of Data and Process Science at the Faculty of Economics and Business, KU Leuven, Belgium. His research interests include business data mining and analytics, machine learning, process management, and process mining. His work has been published in well-known international journals and presented at top conferences. Seppe's teaching includes Advanced Analytics, Big Data, and Information Management courses. He also frequently teaches for industry and business audiences. Besides work, Seppe enjoys traveling, reading (Murakami to Bukowski to Asimov), listening to music (Booka Shade to Miles Davis to Claude Debussy), watching movies and series (less so these days due to a lack of time), gaming, and keeping up with the news.

Bart Baesens is a Professor of Big Data and Analytics at KU Leuven, Belgium, and a lecturer at the University of Southampton, United Kingdom. He has done extensive research on big data and analytics, credit risk modeling, fraud detection, and marketing analytics. Bart has written more than 200 scientific papers and several books. Besides enjoying time with his family, he is also a diehard Club Brugge soccer fan. Bart is a foodie and amateur cook. He loves drinking a good glass of wine (his favorites are white Viognier or red Cabernet Sauvignon) either in his wine cellar or when overlooking the authentic red English phone booth in his garden. Bart loves traveling and is fascinated by World War I and reads many books on the topic.

About the Technical Reviewer

Mark Furman, MBA, is a Systems Engineer, Author, Teacher, and Entrepreneur. For the last 16 years he has worked in the Information Technology field with a focus on Linux-based systems and programming in Python, working for a range of companies including Host Gator, Interland, Suntrust Bank, AT&T, and Winn-Dixie. Currently he has been focusing his career on the maker movement and has launched Tech Forge (techforge. org), which will focus on helping people start a makerspace and help sustain current spaces. He holds a Master of Business Administration from Ohio University. You can follow him on Twitter @mfurman.

Introduction

Congratulations! By picking up this book, you've set the first steps into the exciting world of web scraping. First of all, we want to thank you, the reader, for choosing this guide to accompany you on this journey.

Goals

For those who are not familiar with programming or the deeper workings of the web, web scraping often looks like a black art: the ability to write a program that sets off on its own to explore the Internet and collect data is seen as a magical, exciting, perhaps even scary power to possess. Indeed, there are not many programming tasks that are able to fascinate both experienced and novice programmers in quite such a way as web scraping. Seeing a program working for the first time as it reaches out on the web and starts gathering data never fails to provide a certain rush, feeling like you've circumvented the "normal way" of working and just cracked some sort of enigma. It is perhaps because of this reason that web scraping is also making a lot of headlines these days.

In this book, we set out to provide a concise and modern guide to web scraping, using Python as our programming language. We know that there are a lot of other books and online tutorials out there, but we felt that there was room for another entry. In particular, we wanted to provide a guide that is "short and sweet," without falling into the typical "learn this in X hours" trap where important details or best practices are glossed over just for the sake of speed. In addition, you'll note that we have titled this book as "Practical Web Scraping for Data Science." We're data scientists ourselves, and have very often found web scraping to be a powerful tool to have in your arsenal for the purpose of data gathering. Many data science projects start with the first step of obtaining an appropriate data set. In some cases (the "ideal situation," if you will), a data set is readily provided by a business partner, your company's data warehouse, or your academic supervisor, or can be bought or obtained in a structured format by external data providers; but many truly interesting projects start from collecting a treasure trove of information from the same place as humans do: the web. As such, we set out to offer something that:

- Is concise and to the point, while still being thorough.

- Is geared toward data scientists: we'll show you how web scraping "plugs into" several parts of the data science workflow.

- Takes a code-first approach to get you up to speed quickly without too much boilerplate text.

- Is modern by using well-established best practices and publicly available, open source Python libraries only.

- Goes further than simple basics by showing how to handle the web of today, including JavaScript, cookies, and common web scraping mitigation techniques.

- Includes a thorough managerial and legal discussion regarding web scraping.

- Provides lots of pointers for further reading and learning.

- Includes many larger, fully worked-out examples.

We hope you enjoy reading this book as much as we had writing it. Feel free to contact us in case you have questions, find mistakes, or just want to get in touch! We love hearing from our readers and are open to receive any thoughts and questions.

—Seppe vanden Broucke, seppe.vandenbroucke@kuleuven.be

—Bart Baesens, bart.baesens@kuleuven.be

Audience

We have written this book with a data science-oriented audience in mind. As such, you'll probably already be familiar with Python or some other programming language or analytical toolkit (be it R, SAS, SPSS, or something else). If you're using Python already, you'll feel right at home. If not, we include a quick Python primer later on in this chapter to catch up with the basics and provide pointers to other guides as well. Even if you're not using Python yet for your daily data science tasks (many will argue that you should), we want to show you that Python is a particularly powerful language to use for scraping data from the web. We also assume that you have some basic knowledge regarding how the web works. That is, you know your way around a web browser and know what URLs are; we'll explain the details in depth as we go along.

To summarize, we have written this book to be useful to the following target groups:

- Data science practitioners already using Python and wanting to learn how to scrape the web using this language.

- Data science practitioners using another programming language or toolkit, but want to adopt Python to perform the web scraping part of their pipeline.

- Lecturers and instructors of web scraping courses.

- Students working on a web scraping project or aiming to increase their Python skill set.

- "Citizen data scientists" with interesting ideas requiring data from the web.

- Data science or business intelligence managers wanting to get an overview of what web scraping is all about and how it can bring a benefit to their teams, and what the managerial and legal aspects are that need to be considered.

Structure

The chapters in this book can be divided into three parts:

- **Part I: Web Scraping Basics (Chapters 1-3):** In these chapters, we'll introduce you to web scraping, why it is useful to data scientists, and discuss the key components of the web — HTTP, HTML, and CSS. We'll show you how to write basic scrapers using Python, using the "requests" and "Beautiful Soup" libraries.

- **Part II: Advanced Web Scraping (Chapters 4-6):** Here, we delve deeper into HTTP and show you how to work with forms, login screens, and cookies. We'll also explain how to deal with JavaScript-heavy websites and show you how to go from simple web scrapers to advanced web crawlers.

- **Part III: Managerial Concerns and Best Practices (Chapters 7-9):**
  In this concluding part, we discuss managerial and legal concerns
  regarding web scraping in the context of data science, and also "open
  the door" to explore other tools and interesting libraries. We also list a
  general overview regarding web scraping best practices and tips. The
  final chapter includes some larger web scraping examples to show
  how all concepts covered before can be combined and highlights some
  interesting data science-oriented use cases using web scraped data.

This book is set up to be very easy to read and work through. Newcomers are hence
simply advised to read through this book from start to finish. That said, the book is
structured in such a way that it should be easy to refer back to any part later on in case
you want to brush up your knowledge or look up a particular concept.

PART I

Web Scraping Basics

CHAPTER 1

Introduction

In this chapter, we introduce you to the concept of web scraping and highlight why the practice is useful to data scientists. After illustrating some interesting recent use cases of web scraping across various fields and industry sectors, we make sure you're ready to get started with web scraping by preparing your programming environment.

1.1 What Is Web Scraping?

Web "scraping" (also called "web harvesting," "web data extraction," or even "web data mining"), can be defined as "the construction of an agent to download, parse, and organize data from the web in an automated manner." Or, in other words: instead of a human end user clicking away in a web browser and copy-pasting interesting parts into, say, a spreadsheet, web scraping offloads this task to a computer program that can execute it much faster, and more correctly, than a human can.

The automated gathering of data from the Internet is probably as old as the Internet itself, and the term "scraping" has been around for much longer than the web. Before "web scraping" became popularized as a term, a practice known as "screen scraping" was already well-established as a way to extract data from a visual representation — which in the early days of computing (think 1960s-80s) often boiled down to simple, text-based "terminals." Just as today, people in those days were also interested in "scraping" large amounts of text from such terminals and storing this data for later use.

© Seppe vanden Broucke and Bart Baesens 2018
S. vanden Broucke and B. Baesens, *Practical Web Scraping for Data Science*,
https://doi.org/10.1007/978-1-4842-3582-9_1

1.1.1 Why Web Scraping for Data Science?

When surfing the web using a normal web browser, you've probably encountered multiple sites where you considered the possibility of gathering, storing, and analyzing the data presented on the site's pages. Especially for data scientists, whose "raw material" is data, the web exposes a lot of interesting opportunities:

- There might be an interesting table on a Wikipedia page (or pages) you want to retrieve to perform some statistical analysis.

- Perhaps you want to get a list of reviews from a movie site to perform text mining, create a recommendation engine, or build a predictive model to spot fake reviews.

- You might wish to get a listing of properties on a real-estate site to build an appealing geo-visualization.

- You'd like to gather additional features to enrich your data set based on information found on the web, say, weather information to forecast, for example, soft drink sales.

- You might be wondering about doing social network analytics using profile data found on a web forum.

- It might be interesting to monitor a news site for trending new stories on a particular topic of interest.

The web contains lots of interesting data sources that provide a treasure trove for all sorts of interesting things. Sadly, the current unstructured nature of the web does not always make it easy to gather or export this data in an easy manner. Web browsers are very good at showing images, displaying animations, and laying out websites in a way that is visually appealing to humans, but they do not expose a simple way to export their data, at least not in most cases. Instead of viewing the web page by page through your web browser's window, wouldn't it be nice to be able to automatically gather a rich data set? This is exactly where web scraping enters the picture.

If you know your way around the web a bit, you'll probably be wondering: "Isn't this exactly what Application Programming Interface (APIs) are for?" Indeed, many websites nowadays provide such an API that provides a means for the outside world to access their data repository in a structured way — meant to be consumed and accessed by computer programs, not humans (although the programs are written by humans, of course). Twitter, Facebook, LinkedIn, and Google, for instance, all provide such APIs in order to search and post tweets, get a list of your friends and their likes, see who you're connected with, and so on. So why, then, would we still need web scraping? The point is that APIs are great means to access data sources, provided the website at hand provides one to begin with and that the API exposes the functionality you want. The general rule of thumb is to look for an API first and use that if you can, before setting off to build a web scraper to gather the data. For instance, you can easily use Twitter's API to get a list of recent tweets, instead of reinventing the wheel yourself. Nevertheless, there are still various reasons why web scraping might be preferable over the use of an API:

- The website you want to extract data from does not provide an API.

- The API provided is not free (whereas the website is).

- The API provided is rate limited: meaning you can only access it a number of certain times per second, per day, ...

- The API does not expose all the data you wish to obtain (whereas the website does).

In all of these cases, the usage of web scraping might come in handy. The fact remains that if you can view some data in your web browser, you will be able to access and retrieve it through a program. If you can access it through a program, the data can be stored, cleaned, and used in any way.

1.1.2 Who Is Using Web Scraping?

There are many practical applications of having access to and gathering data on the web, many of which fall in the realm of data science. The following list outlines some interesting real-life use cases:

- Many of Google's products have benefited from Google's core business of crawling the web. Google Translate, for instance, utilizes text stored on the web to train and improve itself.

- Scraping is being applied a lot in HR and employee analytics. The San Francisco-based hiQ startup specializes in selling employee analyses by collecting and examining public profile information, for instance, from LinkedIn (who was not happy about this but was so far unable to prevent this practice following a court case; see *https://www.bloomberg.com/news/features/2017-11-15/the-brutal-fight-to-mine-your-data-and-sell-it-to-your-boss*).

- Digital marketeers and digital artists often use data from the web for all sorts of interesting and creative projects. "We Feel Fine" by Jonathan Harris and Sep Kamvar, for instance, scraped various blog sites for phrases starting with "I feel," the results of which could then visualize how the world was feeling throughout the day.

- In another study, messages scraped from Twitter, blogs, and other social media were scraped to construct a data set that was used to build a predictive model toward identifying patterns of depression and suicidal thoughts. This might be an invaluable tool for aid providers, though of course it warrants a thorough consideration of privacy related issues as well (see *https://www.sas.com/en_ca/insights/articles/analytics/using-big-data-to-predict-suicide-risk-canada.html*).

- Emmanuel Sales also scraped Twitter, though here with the goal to make sense of his own social circle and time line of posts (see *https://emsal.me/blog/4*). An interesting observation here is that the author first considered using Twitter's API, but found that "Twitter heavily rate limits doing this: if you want to get a user's follow list, then you can only do so 15 times every 15 minutes, which is pretty unwieldy to work with."

- In a paper titled "The Billion Prices Project: Using Online Prices for Measurement and Research" (see *http://www.nber.org/papers/w22111*), web scraping was used to collect a data set of online price information that was used to construct a robust daily price index for multiple countries.

- Banks and other financial institutions are using web scraping for competitor analysis. For example, banks frequently scrape competitors' sites to get an idea of where branches are being opened or closed, or to track loan rates offered — all of which is interesting information that can be incorporated in their internal models and forecasting. Investment firms also often use web scraping, for instance, to keep track of news articles regarding assets in their portfolio.

- Sociopolitical scientists are scraping social websites to track population sentiment and political orientation. A famous article called "Dissecting Trump's Most Rabid Online Following" (see *https://fivethirtyeight.com/features/dissecting-trumps-most-rabid-online-following/*) analyzes user discussions on Reddit using semantic analysis to characterize the online followers and fans of Donald Trump.

- One researcher was able to train a deep learning model based on scraped images from Tinder and Instagram together with their "likes" to predict whether an image would be deemed "attractive" (see *http://karpathy.github.io/2015/10/25/selfie/*). Smartphone makers are already incorporating such models in their photo apps to help you brush up your pictures.

- In "The Girl with the Brick Earring," Lucas Woltmann sets out to scrape Lego brick information from *https://www.bricklink.com* to determine the best selection of Lego pieces (see *http://lucaswoltmann.de/art'n'images/2017/04/08/the-girl-with-the-brick-earring.html*) to represent an image (one of the co-authors of this book is an avid Lego fan, so we had to include this example).

- In "Analyzing 1000+ Greek Wines With Python," Florents Tselai scrapes information about a thousand wine varieties from a Greek wine shop (see *https://tselai.com/greek-wines-analysis.html*) to analyze their origin, rating, type, and strength (one of the co-authors of this book is an avid wine fan, so we had to include this example).

- Lyst, a London-based online fashion marketplace, scraped the web for semi-structured information about fashion products and then applied machine learning to present this information cleanly and elegantly for consumers from one central website. Other data scientists have done similar projects to cluster similar fashion products (see *http://talks.lystit.com/dsl-scraping-presentation/*).

- We've supervised a study where web scraping was used to extract information from job sites, to get an idea regarding the popularity of different data science- and analytics-related tools in the workplace (spoiler: Python and R were both rising steadily).

- Another study from our research group involved using web scraping to monitor news outlets and web forums to track public sentiment regarding Bitcoin.

No matter your field of interest, there's almost always a use case to improve or enrich your practice based on data. "Data is the new oil," so the common saying goes, and the web has a lot of it.

1.2 Getting Ready

1.2.1 Setting Up

We'll be using Python 3 throughout this book. You can download and install Python 3 for your platform (Windows, Linux, or MacOS) from *https://www.python.org/downloads/*.

Why Python 3 and Not 2?   According to the creators of Python themselves, "Python 2 is legacy, Python 3 is the present and future of the language." Since we strive to offer a modern guide, we have deliberately chosen Python 3 as our working language. That said, there's still a lot of Python 2 code floating around (maybe even in your organization). Most of the concepts and examples provided in this book should work well in Python 2, too, if you add the following import statement in your Python 2 code:

```
from __future__ import absolute_import, division, print_function
```

You'll also need to install "pip," Python's package manager. If you've installed a recent version of Python 3, it will already come with pip installed. It is a good idea, however, to make sure pip is up to date by executing the following command on the command line:

```
python -m pip install -U pip
```

Or (in case you're using Linux or MacOS):

```
pip install -U pip
```

Manually Installing pip   No pip on your system yet? Refer to the following page to install it on your system (under "Installing with get-pip.py"): *https://pip.pypa.io/en/stable/installing/*.

Finally, you might also wish to install a decent text editor on your system to edit Python code files. Python already comes with a bare-bones editor built in (look for "Idle" in your programs menu), but other text editors such as Notepad++, Sublime Text, VSCode, Atom, and others all work well, too.

1.2.2 A Quick Python Primer

We assume you already have some programming experience under your belt, and perhaps are already somewhat familiar with reading and writing Python code. If not, the following overview will get you up to speed quickly.

Python code can be written and executed in two ways:

1. By using the Python interpreter REPL ("read-eval-print-loop"), which provides an interactive session where you can enter Python commands line by line (read), which will be evaluated (eval), showing the results (print). These steps are repeated ("loop") until you close the session.

2. By typing out Python source code in ".py" files and then running them.

For virtually any use case, it is a good idea to work with proper ".py" files and run these, though the Python REPL comes in handy to test a quick idea or experiment with a few lines of code. To start it, just enter "python" in a command-line window and press enter, as Figure 1-1 shows.

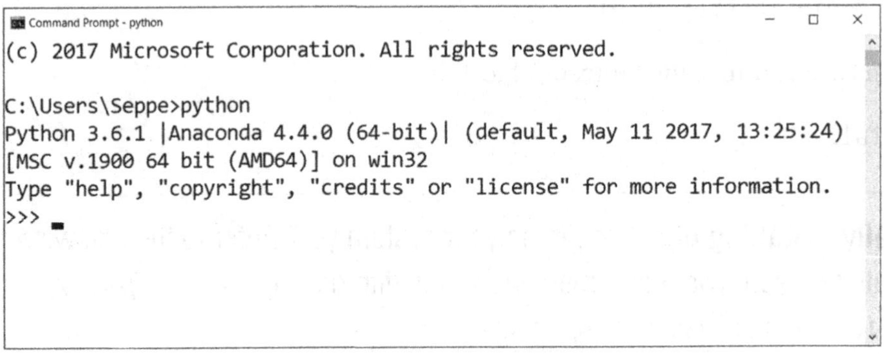

Figure 1-1. *Opening the Python interpreter from the command line*

The three angled brackets ("> > >") indicate a prompt, meaning that the Python REPL is waiting for your commands. Try executing the following commands:

```
>>> 1 + 1
2
>>> 8 - 1
7
>>> 10 * 2
20
>>> 35 / 5
7.0
>>> 5 // 3
1
>>> 7 % 3
1
>>> 2 ** 3
8
```

Mathematics work as you'd expect; "//" indicates an integer division; "%" is the modulo operator, which returns the remainder after division; and "**" means "raised to the power of."

Python supports the following number types:

- Integers ("int"), representing signed integer values (i.e., non-decimal numbers), such as 10, 100, -700, and so on.

- Long integers ("long"), representing signed integer values taking up more memory than standard integers, hence allowing for larger numbers. They're written by putting an "L" at the end, for example, 535633629843L, 10L, -100000000L, and so on.

- Floating-point values ("float"), that is, decimal numbers, such as 0.4, -10.2, and so on.

- Complex numbers ("complex"), which are not widely used in non-mathematics code. They're written by putting a "j" at the end, for example, 3.14j, .876j, and so on.

Apart from numbers, Python also supports strings ("str"): textual values that are enclosed by double or single quotes, as well as the Boolean ("bool") logic values: "True" and "False" (note the capitalization). "None" represents a special value indicating nothingness. Try out the following lines of code to play around with these types:

```
>>> 'This is a string value'
'This is a string value'
>>> "This is also a string value"
'This is also a string value'
>>> "Strings can be " + 'added'
'Strings can be added'
>>> "But not to a number: " + 5
Traceback (most recent call last):
 File "<stdin>", line 1, in <module>
TypeError: must be str, not int
>>> "So we convert it first: " + str(5)
'So we convert it first: 5'
>>> False and False
False
>>> False or True
True
>>> not False
```

```
True
>>> not True
False
>>> not False and True
True
>>> None + 5
Traceback (most recent call last):
  File "<stdin>", line 1, in <module>
TypeError: unsupported operand type(s)
  for +: 'NoneType' and 'int'
>>> 4 < 3 # > also works
False
>>> 100 >= 10 # <= also works
True
>>> 10 == 10
True
>>> None == False
False
>>> False == 0
True
>>> True == 1
True
>>> True == 2
False
>>> 2 != 3
True
>>> 2 != '2'
True
```

Again, the instructions above should be pretty self-explanatory, except for perhaps the last few lines. In Python, "==" indicates an equality comparison and hence returns True or False as a result. None is neither equal to False nor True itself, but False is considered equal to zero (0) and True is considered equal to one (1) in Python. Note that the equality and inequality operators ("==" and "!=") do consider the types that are being compared; the number 2 is hence not equal to the string "2."

Is "Is" Equal to "=="?   Apart from "==", Python also provides the "is" keyword, which will return True if two variables point to the same object (their contents will hence always be equal as well). "==" checks whether the contents of two variables are equal, even though they might not point to the same object. In general, "==" is what you'll want to use, except for a few exceptions, which is to check whether a variable is equal to True, False, or None. All variables having this as their value will point to the same object in memory, so that instead of writing `my_var == None` you can also write `my_var is None` that reads a bit better.

In the REPL interactive session, all results of our inputs are immediately shown on the screen. When executing a Python file, however, this is not the case, and hence we need a function to explicitly "print out" information on the screen. In Python, this can be done through the `print` function:

```
>>> print("Nice to meet you!")
Nice to meet you!
>>> print("Nice", "to", "meet", "you!")
Nice to meet you!
>>> print("HE", "LLO", sep="--")
HE--LLO
>>> print("HELLO", end="!!!\n")
HELLO!!!
```

When working with data, we obviously would like to keep our data around to use in different parts of our program. That is, we'd like to store numbers, strings, ... in variables. Python simply uses the "=" operator for variable assignment:

```
>>> var_a = 3
>>> var_b = 4
>>> var_a + var_b + 2
9
>>> var_str = 'This is a string'
>>> print(var_str)
This is a string
```

Strings in Python can be formatted in a number of different ways. First of all, characters prefixed with a backslash ("\") inside a string indicate so-called "escape characters" and represent special formatting instructions. In the example above, for instance, "\n" indicates a line break. "\t" on the other hand represents a tab, and "\\" is simply the backslash character itself. Next, it is possible to format strings by means of the `format` function:

```
>>> "{} : {}".format("A", "B")
'A : B'
>>> "{0}, {0}, {1}".format("A", "B")
'A, A, B'
>>> "{name} wants to eat {food}".format(name="Seppe", food="lasagna")
'Seppe wants to eat lasagna'
```

Format Overload   If there's anything that new versions of Python don't need, it's more ways to format strings. Apart from using the `format` function illustrated here, Python also allows us to format strings using the "%" operator:

```
"%s is %s" % ("Seppe", "happy")
```

Python 3.6 also added "f-strings" to format strings in a more concise way:

```
f'Her name is {name} and she is {age} years old.'
```

We'll stick to using `format`, to keep things clear.

Other than numbers, Booleans and strings, Python also comes with a number of helpful data structures built in, which we'll be using a lot: lists, tuples, dictionaries, and sets.

Lists are used to store ordered sequences of things. The following instructions outline how they work in Python. Note that the code fragment below also includes comments, which will be ignored by Python and start with a "#" character:

```
>>> li = []
>>> li.append(1) # li is now [1]
>>> li.append(2) # li is now [1, 2]
>>> li.pop() # removes and returns the last element
2
```

```
>>> li = ['a', 2, False] # not all elements need to be the same type
>>> li = [[3], [3, 4], [1, 2, 3]] # even lists of lists
>>> li = [1, 2, 4, 3]
>>> li[0]
1
>>> li[-1]
3
>>> li[1:3]
[2, 4]
>>> li[2:]
[4, 3]
>>> li[:3]
[1, 2, 4]
>>> li[::2] # general format is li[start:end:step]
[1, 4]
>>> li[::-1]
[3, 4, 2, 1]
>>> del li[2] # li is now [1, 2, 3]
>>> li.remove(2) # li is now [1, 3]
>>> li.insert(1, 1000) # li is now [1, 1000, 3]
>>> [1, 2, 3] + [10, 20]
[1, 2, 3, 10, 20]
>>> li = [1, 2, 3]
>>> li.extend([1, 2, 3])
>>> li
[1, 2, 3, 1, 2, 3]
>>> len(li)
6
>>> len('This works for strings too')
26
>>> 1 in li
True
>>> li.index(2)
1
>>> li.index(200)
```

```
Traceback (most recent call last):
  File "<stdin>", line 1, in <module>
ValueError: 200 is not in list
```

Tuples are similar to lists but are immutable, meaning that elements cannot be added or removed after creation:

```
>>> tup = (1, 2, 3)
>>> tup[0]
1
>>> type((1)) # a tuple of length one has to have a comma after the ↵
    last element but tuples of other lengths, even zero, do not
<class 'int'>
>>> type((1,))
<class 'tuple'>
>>> type(())
<class 'tuple'>
>>> len(tup)
3
>>> tup + (4, 5, 6)
(1, 2, 3, 4, 5, 6)
>>> tup[:2]
(1, 2)
>>> 2 in tup
True
>>> a, b, c = (1, 2, 3) # a is now 1, b is now 2 and c is now 3
>>> a, *b, c = (1, 2, 3, 4) # a is now 1, b is now [2, 3] and c is now 4
>>> d, e, f = 4, 5, 6 # you can also leave out the parentheses
>>> e, d = d, e # d is now 5 and e is now 4
```

Sets are also similar to lists, but they store a unique and unordered collection of items, just like a set in mathematics:

```
>>> empty_set = set()
>>> some_set = {1, 1, 2, 2, 3, 4} # some_set is now {1, 2, 3, 4}
>>> filled_set = some_set
>>> filled_set.add(5) # filled_set is now {1, 2, 3, 4, 5}
>>> other_set = {3, 4, 5, 6}
```

```
>>> filled_set & other_set # intersection
{3, 4, 5}
>>> filled_set | other_set # union
{1, 2, 3, 4, 5, 6}
>>> {1, 2, 3, 4} - {2, 3, 5} # difference
{1, 4}
>>> {1, 2} >= {1, 2, 3}
False
>>> {1, 2} <= {1, 2, 3}
True
>>> 2 in filled_set
True
```

Dictionaries store a mapping between a series of unique keys and values:

```
>>> empty_dict = {}
>>> filled_dict = {"one": 1, "two": 2, "three": 3}
>>> filled_dict["one"]
1
>>> list(filled_dict.keys())
["one", "two", "three"]
>>> list(filled_dict.values())
[1, 2, 3]
>>> "one" in filled_dict # in checks based on keys
True
>>> 1 in filled_dict
False
>>> filled_dict.get("one")
1
>>> filled_dict.get("four")
None
>>> filled_dict.get("four", 4) # default value if not found
4
>>> filled_dict.update({"four":4})
>>> filled_dict["four"] = 4 # also possible to add/update this way
>>> del filled_dict["one"] # removes the key "one"
```

Finally, control flow in Python is relatively simple, too:

```
>>> some_var = 10
>>> if some_var > 1:
...     print('Bigger than 1')
...
Bigger than 1
```

Note the colon (":") after the if-statement as well as the three dots "..." in the REPL, indicating that more output is expected before a given piece of code can be executed. Code in Python is structured using white space, meaning that everything inside of an "if"-block, for instance, should be indented using spaces or tabs.

Indentation   Some programmers find this white space indentation frustrating when first working with Python, though it does undeniably lead to more readable and cleanly organized code. Just make sure not to mix tabs and spaces in your source code!

"If"-blocks in Python can also include optional "elif" and "else" blocks:

```
>>> some_var = 10
>>> if some_var > 10:
...     print('Bigger than 10')
... elif some_var > 5:
...     print('Bigger than 5')
... else:
...     print('Smaller than or equal to 5')
...
Bigger than 5
```

Readable If Blocks   Remember that zero (0) integers, floats, and complex numbers all evaluate to False in Python. Similarly, empty strings, sets, tuples, lists, and dictionaries also evaluate to False, so instead of writing `if len(my_list) > 0:`, you can simply use `if my_list:` as well, which is much easier to read.

We've already seen the "in"-operator as away to check for list, tuple, set, and dictionary membership. This operator can be used to write "for"-loops as well:

```
>>> some_list = [1, 2, 3]
>>> some_string = 'a string'
>>> 1 in some_list
True
>>> 'string' in some_string
True
>>> for num in some_list:
... print(num)
...
1
2
3
>>> for chr in some_string:
...     print(chr)
...
a

s
t
r
i
n
g
```

To loop over number ranges, Python also provides the helpful built-in range function:

- range(number): returns an iterable of numbers from zero to (not including) the given number.

- range(lower, upper): returns an iterable of numbers from the lower number to (not including) the upper number.

- range(lower, upper, step): returns an iterable of numbers from the lower number to the upper number, while incrementing by step.

Integers Only   All of these functions sadly require integers as input arguments. If you want to iterate over a range of decimal values, you'll have to define your own function.

Note the use of the concept "iterable" here. In Python, iterables are basically a "smart" list. Instead of immediately filling up your computer's memory with the complete list, Python will avoid doing so until you actually need to access the elements themselves. This is why using the range function shows the following:

```
>>> range(3)
range(0, 3)
```

Converting iterables to a real list is simple; just convert the value to an explicit list:

```
>>> list(range(3))
[0, 1, 2]
```

While looping over an iterable, however, you don't need to explicitly convert them first.

You can hence just use the range functions directly as follows:

```
>>> for num in range(1, 100, 15):
...     print(num)
...
1
16
31
46
61
76
91
```

And, of course, Python has a "while"-style loop as well:

```
>>> x = 0
>>> while x < 3:
...     print(x)
...     x = x + 1
```

20

```
...
0
1
2
```

Infinite Loops   Forgot to add the x = x + 1 line? Tried out while True:, or is your "for"-loop going in circles? You can press Control+C on your keyboard to execute a "Keyboard Interrupt" and stop the execution of the current code block.

Whenever you're writing code, it is a good idea to encapsulate small reusable pieces of code so they can be called and executed at various places without having to copy-paste lots of code. A basic way to do so is to create a function, which is done using "def":

```
>>> def add(x, y):
...     print("x is {} and y is {}".format(x, y))
...     return x + y  # return value
...
>>> add(5, 6)
x is 5 and y is 6
11
>>> add(y=10, x=5)
x is 5 and y is 10
15
```

There are two special constructs worth mentioning here as well: "*" and "**". Both of these can be used in function signatures to indicate "the rest of the arguments" and "the rest of the named arguments" respectively. Let's show how they work using an example:

```
>>> def many_arguments(*args):
...     # args will be a tuple
...     print(args)
...
>>> many_arguments(1, 2, 3)
(1, 2, 3)
>>> def many_named_arguments(**kwargs):
...     # kwargs will be a dictionary
...     print(kwargs)
```

```
...
>>> many_named_arguments(a=1, b=2)
{'a': 1, 'b': 2}
>>> def both_together(*args, **kwargs):
...    print(args, kwargs)
```

Apart from using these in method signatures, you can also use them when calling functions to indicate that you want to pass an iterable as arguments or a dictionary as named arguments:

```
>>> def add(a, b):
...    return a + b
...
>>> l = [1,2] # tuples work, too
>>> add(*l)
3
>>> d = {'a': 2, 'b': 1}
>>> add(**d)
3
```

Finally, instead of using the Python REPL, let's take a look at how you'd write Python code using a source file. Create a file called "test.py" somewhere where you can easily find it and add the following contents:

```
# test.py

def add(x, y):
  return x + y

for i in range(5):
  print(add(i, 10))
```

Save the file, and open a new command-line window. You can execute this file by supplying its name to the "python" executable, as Figure 1-2 shows:

```
Command Prompt                                              —    □    ×
C:\Users\Seppe>cd Desktop

C:\Users\Seppe\Desktop>python test.py
10
11
12
13
14

C:\Users\Seppe\Desktop>
```

Figure 1-2. *Running a Python source file ("test.py") from the command line*

This concludes our "rapid-fire" Python primer. We've skipped over some details here (such as classes, try-except-catch blocks, iterators versus generators, inheritance, and so on), but these are not really necessary to get started with what we're actually here to do: web scraping.

In case you're looking for more resources on Python programming, check out these links:

- The Official Python 3 Documentation: *https://docs.python.org/3/*

- Dive Into Python 3: *http://www.diveintopython3.net/index.html*

- Automate the Boring Stuff with Python: *https://automatetheboringstuff.com/*

- The Hitchhiker's Guide to Python: *http://docs.python-guide.org/en/latest/*

- First Steps With Python: *https://realpython.com/learn/python-first-steps/*

CHAPTER 2

The Web Speaks HTTP

In this chapter, we introduce one of the core building blocks that makes up the web: the HyperText Transfer Protocol (HTTP), after having provided a brief introduction to computer networks in general. We then introduce the Python requests library, which we'll use to perform HTTP requests and effectively start retrieving websites with Python. The chapter closes with a section on using parameters in URLs.

2.1 The Magic of Networking

Nowadays, the web has become so integrated into our day-to-day activities that we rarely consider its complexity. Whenever you surf the web, a whole series of networking protocols is being kicked into gear to set up connections to computers all over the world and retrieve data, all in a matter of seconds. Consider, for instance, the following series of steps that gets executed by your web browser once you navigate to a website, say `www.google.com`:

1. You enter "`www.google.com`" into your web browser, which needs to figure out the IP address for this site. IP stands for "Internet Protocol" and forms a core protocol of the Internet, as it enables networks to route and redirect communication packets between connected computers, which are all given an IP address. To communicate with Google's web server, you need to know its IP address. Since the IP address is basically a number, it would be kind of annoying to remember all these numbers for every website out there. So, just as how you link telephone numbers to names in your phone's contact book, the web provides a mechanism to translate domain names like "`www.google.com`" to an IP address.

© Seppe vanden Broucke and Bart Baesens 2018
S. vanden Broucke and B. Baesens, *Practical Web Scraping for Data Science*,
https://doi.org/10.1007/978-1-4842-3582-9_2

2.   And so, your browser sets off to figure out the correct IP address behind "www.google.com". To do so, your web browser will use another protocol, called DNS (which stands for Domain Name System) as follows: first, the web browser will inspect its own cache (its "short term memory") to see whether you've recently visited this website in the past. If you have, the browser can reuse the stored address. If not, the browser will ask the underlying operating system (Windows, for example) to see whether it knows the address for www.google.com.

3.   If the operating system is also unaware of this domain, the browser will send a DNS request to your router, which is the machine that connects you to the Internet and also — typically — keeps its own DNS cache. If your router is also unaware of the correct address, your browser will start sending a number of data packets to known DNS servers, for example, to the DNS server maintained by your Internet Service Provider (ISP) — for which the IP address is known and stored in your router. The DNS server will then reply with a response basically indicating that "www.google.com" is mapped to the IP address "172.217.17.68". Note that even your ISPs DNS server might have to ask other DNS servers (located higher in the DNS hierarchy) in case it doesn't have the record at hand.

4.   All of this was done just to figure out the IP address of www.google.com. Your browser can now establish a connection to 172.217.17.68, Google's web server. A number of protocols — a protocol is a standard agreement regarding what messages between communicating parties should look like — are combined here (wrapped around each other, if you will) to construct a complex message. At the outermost part of this "onion," we find the IEEE 802.3 (Ethernet) protocol, which is used to communicate with machines on the same network. Since we're not communicating on the same network, the Internet Protocol, IP, is used to embed another message indicating that we wish to contact the server at address 172.217.17.68. Inside this, we find another protocol, called TCP (Transmission Control Protocol),

which provides a general, reliable means to deliver network messages, as it includes functionality for error checking and splitting messages up in smaller packages, thereby ensuring that these packets are delivered in the right order. TCP will also resend packets when they are lost in transmission. Finally, inside the TCP message, we find another message, formatted according to the HTTP protocol (HyperText Transfer Protocol), which is the actual protocol used to request and receive web pages. Basically, the HTTP message here states a request from our web browser: "Can I get your index page, please?"

5. Google's web server now sends back an HTTP reply, containing the contents of the page we want to visit. In most cases, this textual content is formatted using HTML, a markup language we'll take a closer look at later on. From this (oftentimes large) bunch of text, our web browser can set off to render the actual page, that is, making sure that everything appears neatly on screen as instructed by the HTML content. Note that a web page will oftentimes contain pieces of content for which the web browser will — behind the scenes — initiate new HTTP requests. In case the received page instructs the browser to show an image, for example, the browser will fire off another HTTP request to get the contents of the image (which will then not look like HTML-formatted text but simply as raw, binary data). As such, rendering just one web page might involve a large deal of HTTP requests. Luckily, modern browsers are smart and will start rendering the page as soon as information is coming in, showing images and other visuals as they are retrieved. In addition, browsers will try to send out multiple requests in parallel if possible to speed up this process as well.

With so many protocols, requests, and talking between machines going on, it is nothing short of amazing that you are able to view a simple web page in less than a second. To standardize the large amount of protocols that form the web, the

International Organization of Standardization (ISO) maintains the Open Systems Interconnection (OSI) model, which organizes computer communication into seven layers:

- **Layer 1: Physical Layer:** Includes the Ethernet protocol, but also USB, Bluetooth, and other radio protocols.

- **Layer 2: Data link Layer:** Includes the Ethernet protocol.

- **Layer 3: Network Layer:** Includes IP (Internet Protocol).

- **Layer 4: Transport Layer:** TCP, but also protocols such as UDP, which do not offer the advanced error checking and recovery mechanisms of TCP for sake of speed.

- **Layer 5: Session Layer:** Includes protocols for opening/closing and managing sessions.

- **Layer 6: Presentation Layer:** Includes protocols to format and translate data.

- **Layer 7: Application Layer:** HTTP and DNS, for instance.

Not all network communications need to use protocols from all these layers. To request a web page, for instance, layers 1 (physical), 2 (Ethernet), 3 (IP), 4 (TCP), and 7 (HTTP) are involved, but the layers are constructed so that each protocol found at a higher level can be contained inside the message of a lower-layer protocol. When you request a secure web page, for instance, the HTTP message (layer 7) will be encoded in an encrypted message (layer 6) (this is what happens if you surf to an "https"-address). The lower the layer you aim for when programming networked applications, the more functionality and complexity you need to deal with. Luckily for us web scrapers, we're interested in the topmost layer, that is, HTTP, the protocol used to request and receive web pages. That means that we can leave all complexities regarding TCP, IP, Ethernet, and even resolving domain names with DNS up to the Python libraries we use, and the underlying operating system.

2.2 The HyperText Transfer Protocol: HTTP

We've now seen how your web browser communicates with a server on the World Wide Web. The core component in the exchange of messages consists of a HyperText Transfer Protocol (HTTP) request message to a web server, followed by an HTTP response (also oftentimes called an HTTP reply), which can be rendered by the browser.

Since all of our web scraping will build upon HTTP, we do need to take a closer look at HTTP messages to learn what they look like.

HTTP is, in fact, a rather simple networking protocol. It is text based, which at least makes its messages somewhat readable to end users (compared to raw binary messages that have no textual structure at all) and follow a simple request-reply-based communication scheme. That is, contacting a web server and receiving a reply simply involves two HTTP messages: a request and a reply. In case your browser wants to download or fetch additional resources (such as images), this will simply entail additional request-reply messages being sent.

Keep Me Alive   In the simplest case, every request-reply cycle in HTTP involves setting up a fresh new underlying TCP connection as well. For heavy websites, setting up many TCP connections and tearing them down in quick succession creates a lot of overhead, so HTTP version 1.1 allows us to keep the TCP connection "alive" to be used for concurrent request-reply HTTP messages. HTTP version 2.0 even allows us to "multiplex" (a fancy word for "mixing messages") in the same connection, for example, to send multiple concurrent requests. Luckily, we don't need to concern ourselves much with these details while working with Python, as the requests, the library we'll use, takes care of this for us automatically behind the scenes.

Let us now take a look at what an HTTP request and reply look like. As we recall, a client (the web browser, in most cases) and web server will communicate by sending plain text messages. The client sends requests to the server and the server sends responses, or replies.

A request message consists of the following:

- A request line;

- A number of request headers, each on their own line;

- An empty line;

- An optional message body, which can also take up multiple lines.

Each line in an HTTP message must end with <CR><LF> (the ASCII characters 0D and 0A).

The empty line is simply <CR><LF> with no other additional white space.

New Lines   <CR> and <LF> are two special characters to indicate that a new line should be started. You don't see them appearing as such, but when you type out a plain text document in, say, Notepad, every time you press enter, these two characters will be put inside of the contents of the document to represent "that a new line appears here." An annoying aspect of computing is that operating systems do not always agree on which character to use to indicate a new line. Linux programs tend to use <LF> (the "line feed" character), whereas older versions of MacOS used <CR> (the "carriage return" character). Windows uses both <CR> and <LF> to indicate a new line, which was also adopted by the HTTP standard. Don't worry too much about this, as the Python requests library will take care of correctly formatting the HTTP messages for us.

The following code fragment shows a full HTTP request message as executed by a web browser (we don't show the "<CR><LF>" after each line, except for the last, blank line):

```
GET / HTTP/1.1
Host: example.com
Connection: keep-alive
Cache-Control: max-age=0
Upgrade-Insecure-Requests: 1
User-Agent: Mozilla/5.0 (Windows NT 10.0; Win64; x64) AppleWebKit/537.36 ↵
    (KHTML, like Gecko) Chrome/60.0.3112.90 Safari/537.36
Accept: text/html,application/xhtml+xml,application/xml;q=0.9,*/*;q=0.8
Referer: https://www.google.com/
Accept-Encoding: gzip, deflate
Accept-Language: en-US,en;q=0.8,nl;q=0.6
<CR><LF>
```

Let's take a closer look at this message. "GET / HTTP/1.1" is the request line. It contains the HTTP "verb" or "method" we want to execute ("GET" in the example above), the URL we want to retrieve ("/"), and the HTTP version we understand ("HTTP/1.1"). Don't worry too much about the "GET" verb. HTTP has a number of verbs (that we'll discuss later on). For now, it is important to know that "GET" means this: "get the contents of this URL for me." Every time you enter a URL in your address bar and press enter, your browser will perform a GET request.

Next up are the request headers, each on their own line. In this example, we already have quite a few of them. Note that each header includes a name ("Host," for instance), followed by a colon (":") and the actual value of the header ("example.com"). Browsers are very chatty in terms of what they like to include in their headers, and Chrome (the web browser used here, is no exception).

The HTTP standard includes some headers that are standardized and which will be utilized by proper web browsers, though you are free to include additional headers as well. "Host," for instance, is a standardized and mandatory header in HTTP 1.1 and higher. The reason why it was not around in HTTP 1.0 (the first version) is simple: in those days, each web server (with its IP address) was responsible for serving one particular website. If we would hence send "GET / HTTP/1.1" to a web server responsible for "example.com", the server knew which page to fetch and return. However, it didn't take long for the following bright idea to appear: Why not serve multiple websites from the same server, with the same IP address? The same server responsible for "example.com" might also be the one serving pages belonging to "example.org", for instance. However, we then need a way to tell the server which domain name we'd like to retrieve a page from. Including the domain name in the request line itself, like "GET example.org/ HTTP/1.1" might have been a solid idea, though this would break backward compatibility with earlier web servers, which expect a URL without a domain name in the request line. A solution was then offered in the form of a mandatory "Host" header, indicating from which domain name the server should retrieve the page.

Wrong Host   Don't try to be too clever and send a request to a web server responsible for "example.com" and change the "Host" header to read "Host: somethingentirely-different.com". Proper web servers will complain and simply send back an error page saying: "Hey, I'm not the server hosting that domain." This being said, security issues have been identified on websites where it is possible to confuse and misdirect them by spoofing this header.

Apart from the mandatory "Host" header, we also see a number of other headers appearing that form a set of "standardized requests headers," which are not mandatory, though nevertheless included by all modern web browsers. "Connection: keep-alive," for instance, signposts to the server that it should keep the connection open for subsequent requests if it can. The "User-Agent" contains a large text value through which the browser happily informs the server what it is (Chrome), and which version it is running as.

The User-Agent Mess   Well… you'll note that the "User-Agent" header contains "Chrome," but also a lot of additional seemingly unrelated text such as "Mozilla," "AppleWebKit," and so on. Is Chrome masquerading itself and posing as other browsers? In a way, it it, though it is not the only browser that does so. The problem is this: when the "User-Agent" header came along and browsers started sending their names and version, some website owners thought it was a good idea to check this header and reply with different versions of a page depending on who's asking, for instance to tell users that "Netscape 4.0" is not supported by this server. The routines responsible for these checks were often implemented in a haphazardly way, thereby mistakenly sending users off when they're running some unknown browser, or failing to correctly check the browser's version. Browser vendors hence had no choice over the years to get creative and include lots of other text fields in this User-Agent header. Basically, our browser is saying "I'm Chrome, but I'm also compatible with all these other browsers, so just let me through please."

"Accept" tells the server which forms of content the browser prefers to get back, and "Accept-Encoding" tells the server that the browser is also able to get back compressed content. The "Referer" header (a deliberate misspelling) tells the server from which page the browser comes from (in this case, a link was clicked on "google.com" sending the browser to "example.com").

A Polite Request   Even though your web browser will try to behave politely and, for instance, tell the web server which forms of content it accepts, there is no guarantee whatsoever that a web server will actually look at these headers or follow up on them. A browser might indicate in its "Accept" header that it understands "webp" images, but the web server can just ignore this request and send back images as "jpg" or "png" anyway. Consider these request headers as polite requests, though, nothing more.

Finally, our request message ends with a blank <CR><LF> line, and has no message body whatsoever. These are not included in GET requests, but we'll see HTTP messages later on where this message body will come into play.

If all goes well, the web server will process our request and send back an HTTP reply. These look very similar to HTTP requests and contain:

- A status line that includes the status code and a status message;

- A number of response headers, again all on the same line;

- An empty line;

- An optional message body.

As such, we might get the following response following our request above:

```
HTTP/1.1 200 OK
Connection: keep-alive
Content-Encoding: gzip
Content-Type: text/html;charset=utf-8
Date: Mon, 28 Aug 2017 10:57:42 GMT
Server: Apache v1.3
Vary: Accept-Encoding
Transfer-Encoding: chunked
<CR><LF>
<html>
<body>Welcome to My Web Page</body>
</html>
```

Again, let's take a look at the HTTP reply line by line. The first line indicates the status result of the request. It opens by listing the HTTP version the server understands ("HTTP/1.1"), followed by a status code ("200"), and a status message ("OK"). If all goes well, the status will be 200. There are a number of agreed-upon HTTP status codes that we'll take a closer look at later on, but you're probably also familiar with the 404 status message, indicating that the URL listed in the request could not be retrieved, that was, was "not found" on the server.

Next up are — again — a number of headers, now coming from the server. Just like web browsers, servers can be quite chatty in terms of what they provide, and can include as many headers as they like. Here, the server includes its current date and version

("Apache v1.3") in its headers. Another important header here is "Content-Type," as it will provide browsers with information regarding what the content included in the reply looks like. Here, it is HTML text, but it might also be binary image data, movie data, and so on.

Following the headers is a blank <CR><LF> line, and an optional message body, containing the actual content of the reply. Here, the content is a bunch of HTML text containing "Welcome to My Web Page." It is this HTML content that will then be parsed by your web browser and visualized on the screen. Again, the message body is optional, but since we expect most requests to actually come back with some content, a message body will be present in almost all cases.

Message Bodies   Even when the status code of the reply is 404, for instance, many websites will include a message body to provide the user with a nice looking page indicating that — sorry — this page could not be found. If the server leaves it out, the web browser will just show its default "Page not found" page instead. There are some other cases where an HTTP reply does not include a message body, which we'll touch upon later on.

2.3 HTTP in Python: The Requests Library

We've now seen the basics regarding HTTP, so it is time we get our hands dirty with some Python code. Recall the main purpose of web scraping: to retrieve data from the web in an automated manner. Basically, we're throwing out our web browser and we're going to surf the web using a Python program instead. This means that our Python program will need to be able to speak and understand HTTP.

Definitely, we could try to program this ourselves on top of standard networking functionality already built-in in Python (or other languages, for that manner), making sure that we neatly format HTTP request messages and are able to parse the incoming responses. However, we're not interested in reinventing the wheel, and there are many Python libraries out there already that make this task a lot more pleasant, so that we can focus on what we're actually trying to accomplish.

In fact, there are quite a few libraries in the Python ecosystem that can take care of HTTP for us. To name a few:

- Python 3 comes with a built-in module called "urllib," which can deal with all things HTTP (see *https://docs.python.org/3/library/urllib.html*). The module got heavily revised compared to its counterpart in Python 2, where HTTP functionality was split up in both "urllib" and "urllib2" and somewhat cumbersome to work with.

- "httplib2" (see *https://github.com/httplib2/httplib2*): a small, fast HTTP client library. Originally developed by Googler Joe Gregorio, and now community supported.

- "urllib3" (see *https://urllib3.readthedocs.io/*): a powerful HTTP client for Python, used by the requests library below.

- "requests" (see *http://docs.python-requests.org/*): an elegant and simple HTTP library for Python, built "for human beings."

- "grequests" (see *https://pypi.python.org/pypi/grequests*): which extends requests to deal with asynchronous, concurrent HTTP requests.

- "aiohttp" (see *http://aiohttp.readthedocs.io/*): another library focusing on asynchronous HTTP.

In this book, we'll use the "requests" library to deal with HTTP. The reason why is simple: whereas "urllib" provides solid HTTP functionality (especially compared with the situation in Python 2), using it often involves lots of boilerplate code making the module less pleasant to use and not very elegant to read. Compared with "urllib," "urllib3" (not part of the standard Python modules) extends the Python ecosystem regarding HTTP with some advanced features, but it also doesn't really focus that much on being elegant or concise. That's where "requests" comes in. This library builds on top of "urllib3," but it allows you to tackle the majority of HTTP use cases in code that is short, pretty, and easy to use. Both "grequests" and "aiohttp" are more modern-oriented libraries and aim to make HTTP with Python more asynchronous. This becomes especially important for very heavy-duty applications where you'd have to make lots of HTTP requests as quickly as possible. We'll stick with "requests" in what follows, as asynchronous programming is a rather challenging topic on its own, and we'll discuss more traditional ways of speeding up your web scraping programs in a robust manner.

It should not be too hard to move on from "requests" to "grequests" or "aiohttp" (or other libraries) should you wish to do so later on.

Installing requests can be done easily through pip (refer back to section 1.2.1 if you still need to set up Python 3 and pip). Execute the following in a command-line window (the "-U" argument will make sure to update an existing version of requests should there already be one):

```
pip install -U requests
```

Next, create a Python file ("firstexample.py" is a good name), and enter the following:

```
import requests

url = 'http://www.webscrapingfordatascience.com/basichttp/'
r = requests.get(url)
print(r.text)
```

If all goes well, you should see the following line appear when executing this script:

```
Hello from the web!
```

Webscrapingfordatascience Dot Com?   *www.webscrapingfordatascience.com* is the companion website for this book. We'll use the pages on this site throughout this book to show off various examples. Since the web is a fast-moving place, we wanted to make sure that the examples we provide continue working as long as possible. Don't be too upset about staying in the "safe playground" for now, as various real-life examples are included in the last chapter as well.

Let's take a look at what's happening in this short example:

- First, we import the requests module. If you've installed requests correctly on your system, the import line should simply work without any errors or warnings.

- We're going to retrieve the contents of *http://www.webscrapingfor datascience.com/basichttp/*. Try opening this web page in your browser. You'll see "Hello from the web!" appear on the page. This is what we want to extract using Python.

- We use the `requests.get` method to perform an "HTTP GET" request to the provided URL. In the simplest case, we only need to provide the URL of the page we want to retrieve. Requests will make sure to format a proper HTTP request message in accordance with what we've seen before.

- The `requests.get` method returns a `requests.Response` Python object containing lots of information regarding the HTTP reply that was retrieved. Again, requests takes care of parsing the HTTP reply so that you can immediately start working with it.

- `r.text` contains the HTTP response content body in a textual form. Here, the HTTP response body simple contained the content "Hello from the web!"

A More Generic Request   Since we'll be working with HTTP GET requests only (for now), the `requests.get` method will form a cornerstone of the upcoming examples. Later on, we'll also deal with other types of HTTP requests, such as POST. Each of these come with a corresponding method in requests, for example, `requests.post`. There's also a generic request method that looks like this: `requests.request('GET', url)`. This is a bit longer to write, but might come in handy in cases where you don't know beforehand which type of HTTP request (GET, or something else) you're going to make.

Let us expand upon this example a bit further to see what's going on under the hood:

```
import requests

url = 'http://www.webscrapingfordatascience.com/basichttp/'
r = requests.get(url)

# Which HTTP status code did we get back from the server?
print(r.status_code)
# What is the textual status code?
print(r.reason)
# What were the HTTP response headers?
print(r.headers)
```

```
# The request information is saved as a Python object in r.request:
print(r.request)
# What were the HTTP request headers?
print(r.request.headers)

# The HTTP response content:
print(r.text)
```

If you run this code, you'll see the following result:

```
200
OK
{'Date': 'Wed, 04 Oct 2017 08:26:03 GMT',
 'Server': 'Apache/2.4.18 (Ubuntu)',
 'Content-Length': '20',
 'Keep-Alive': 'timeout=5, max=99',
 'Connection': 'Keep-Alive',
 'Content-Type': 'text/html; charset=UTF-8'}
<PreparedRequest [GET]>
{'User-Agent': 'python-requests/2.18.4',
 'Accept-Encoding': 'gzip, deflate',
 'Accept': '*/*',
 'Connection': 'keep-alive'}
Hello from the web!
```

- Recall our earlier discussion on HTTP requests and replies. By using the status_code and reason attributes of a request.Response object, we can retrieve the HTTP status code and associated text message we got back from the server. Here, a status code and message of "200 OK" indicates that everything went well.

- The headers attribute of the request.Response object returns a dictionary of the headers the server included in its HTTP reply. Again: servers can be pretty chatty. This server reports its data, server version, and also provides the "Content-Type" header.

- To get information regarding the HTTP request that was fired off, you can access the `request` attribute of a `request.Response` object. This attribute itself is a `request.Request` object, containing information about the HTTP request that was prepared.

- Since an HTTP request message also includes headers, we can access the `headers` attribute for this object as well to get a dictionary representing the headers that were included by requests. Note that requests politely reports its "User-Agent" by default. In addition, requests can take care of compressed pages automatically as well, so it also includes an "Accept-Encoding" header to signpost this. Finally, it includes an "Accept" header to indicate that "any format you have can be sent back" and can deal with "keep-alive" connections as well. Later on, however, we'll see cases where we need to override requests' default request header behavior.

2.4 Query Strings: URLs with Parameters

There's one more thing we need to discuss regarding the basic working of HTTP: URL parameters. Try adapting the code example above in order to scrape the URL *http://www.webscrapingfordatascience.com/paramhttp/*. You should get the following content:

```
Please provide a "query" parameter
```

Try opening this page in your web browser to verify that you get the same result. Now try navigating to the page *http://www.webscrapingfordatascience.com/paramhttp/?query=test*. What do you see?

The optional "?..." part in URLs is called the "query string," and it is meant to contain data that does not fit within a URL's normal hierarchical path structure. You've probably encountered this sort of URL many times when surfing the web, for example:

- `http://www.example.com/product_page.html?product_id=304`

- `https://www.google.com/search?dcr=0&source=hp&q=test&oq=test`

- `http://example.com/path/to/page/?type=animal&location=asia`

Web servers are smart pieces of software. When a server receives an HTTP request for such URLs, it may run a program that uses the parameters included in the query string — the "URL parameters" — to render different content. Compare *http://www.webscrapingfordatascience.com/paramhttp/?query=test* with *http://www.webscrapingfordatascience.com/paramhttp/?query=anothertest*, for instance. Even for this simple page, you see how the response dynamically incorporates the parameter data that you provided in the URL.

Query strings in URLs should adhere to the following conventions:

- A query string comes at the end of a URL, starting with a single question mark, "?".

- Parameters are provided as key-value pairs and separated by an ampersand, "&".

- The key and value are separated using an equals sign, "=".

- Since some characters cannot be part of a URL or have a special meaning (the characters "/", "?", "&", and "=" for instance), URL "encoding" needs to be applied to properly format such characters when using them inside of a URL. Try this out using the URL *http://www.webscrapingfordatascience.com/paramhttp/?query=another%20test%3F%26*, which sends "another test?&" as the value for the "query" parameter to the server in an encoded form.

- Other exact semantics are not standardized. In general, the order in which the URL parameters are specified is not taken into account by web servers, though some might. Many web servers will also be able to deal and use pages with URL parameters without a value, for example, *http://www.example.com/?noparam=&anotherparam*. Since the full URL is included in the request line of an HTTP request, the web server can decide how to parse and deal with these.

URL Rewriting   This latter remark also highlights another important aspect regarding URL parameters: even although they are somewhat standardized, they're not treated as being a "special" part of a URL, which is just sent as a plain text line in an HTTP request anyway. Most web servers will pay attention to parse them on their end in order to use their information while rendering a page (or even ignore them when they're unused — try the URL *http://www. webscrapingfordatascience.com/paramhttp/?query=test&other= ignored*, for instance), but in recent years, the usage of URL parameters is being avoided somewhat. Instead, most web frameworks will allow us to define "nice looking" URLs that just include the parameters in the path of a URL, for example, "/product/302/" instead of "products.html?p=302". The former looks nicer when looking at the URL as a human, and search engine optimization (SEO) people will also tell you that search engines prefer such URLs as well. On the server-side of things, any incoming URL can hence be parsed at will, taking pieces from it and "rewriting" it, as it is called, so some parts might end up being used as input while preparing a reply. For us web scrapers, this basically means that even although you don't see a query string in a URL, there might still be dynamic parts in the URL to which the server might respond in different ways.

Let's take a look at how to deal with URL parameters in requests. The easiest way to deal with these is to include them simply in the URL itself:

```
import requests

url = 'http://www.webscrapingfordatascience.com/paramhttp/?query=test'
r = requests.get(url)

print(r.text)
# Will show: I don't have any information on "test"
```

In some circumstances, requests will try to help you out and encode some characters for you:

```
import requests

url = 'http://www.webscrapingfordatascience.com/paramhttp/?query=a query
with spaces'
r = requests.get(url)
# Parameter will be encoded as 'a%20query%20with%20spaces'
# You can verify this be looking at the prepared request URL:
print(r.request.url)
# Will show [...]/paramhttp/?query=a%20query%20with%20spaces

print(r.text)
# Will show: I don't have any information on "a query with spaces"
```

However, sometimes the URL is too ambiguous for requests to make sense of it:

```
import requests

url = 'http://www.webscrapingfordatascience.com/paramhttp/?query=complex?&'

# Parameter will not be encoded
r = requests.get(url)

# You can verify this be looking at the prepared request URL:
print(r.request.url)
# Will show [...]/paramhttp/?query=complex?&

print(r.text)
# Will show: I don't have any information on "complex?"
```

In this case, requests is unsure whether you meant "?&" to belong to the actual URL as is or whether you wanted to encode it. Hence, requests will do nothing and just request the URL as is. On the server-side, this particular web server is able to derive that the second question mark ("?") should be part of the URL parameter (and should have been properly encoded, but it won't complain), though the ampersand "&" is too ambiguous in this case. Here, the web server assumes that it is a normal separator and not part of the URL parameter value.

So how then, can we properly resolve this issue? A first method is to use the "url-lib.parse" functions `quote` and `quote_plus`. The former is meant to encode special characters in the path section of URLs and encodes special characters using percent "%XX" encoding, including spaces. The latter does the same, but replaces spaces by plus signs, and it is generally used to encode query strings:

```
import requests
from urllib.parse import quote, quote_plus

raw_string = 'a query with /, spaces and?&'
print(quote(raw_string))
print(quote_plus(raw_string))
```

This example will print out these two lines:

```
a%20query%20with%20/%2C%20spaces%20and%3F%26
a+query+with+%2F%2C+spaces+and%3F%26
```

The `quote` function applies percent encoding, but leaves the slash ("/") intact (as its default setting, at least) as this function is meant to be used on URL paths. The `quote_plus` function does apply a similar encoding, but uses a plus sign ("+") to encode spaces and will also encode slashes. As long as we make sure that our query parameter does not use slashes, both encoding approaches are valid to be used to encode query strings. In case our query string does include a slash, and if we do want to use `quote`, we can simply override its `safe` argument as done below:

```
import requests
from urllib.parse import quote, quote_plus

raw_string = 'a query with /, spaces and?&'
url = 'http://www.webscrapingfordatascience.com/paramhttp/?query='

print('\nUsing quote:')
# Nothing is safe, not even '/' characters, so encode everything
r = requests.get(url + quote(raw_string, safe=''))
print(r.url)
print(r.text)
```

```
print('\nUsing quote_plus:')
r = requests.get(url + quote_plus(raw_string))
print(r.url)
print(r.text)
```

This example will print out:

```
Using quote:
http://[...]/?query=a%20query%20with%20%2F%2C%20spaces%20and%3F%26
I don't have any information on "a query with /, spaces and?&"
```

```
Using quote_plus:
http://[...]/?query=a+query+with+%2F%2C+spaces+and%3F%26
I don't have any information on "a query with /, spaces and?&"
```

All this encoding juggling can quickly lead to a headache. Wasn't requests supposed to make our life easy and deal with this for us? Not to worry, as we can simply rewrite the example above using requests only as follows:

```
import requests

url = 'http://www.webscrapingfordatascience.com/paramhttp/'

parameters = {
    'query': 'a query with /, spaces and?&'
    }

r = requests.get(url, params=parameters)
```

```
print(r.url)
print(r.text)
```

Note the usage of the params argument in the requests.get method: you can simply pass a Python dictionary with your non-encoded URL parameters and requests will take care of encoding them for you.

Empty and Ordered Parameters   Empty parameters, for example, as in "params={'query': ''}" will end up in the URL with an equals sign included, that is, "?query=". If you want, you can also pass a list to `params` with every element being a tuple or list itself having two elements representing the key and value per parameter respectively, in which case the order of the list will be respected. You can also pass an `OrderedDict` object (a built-in object provided by the "collections" module in Python 3) that will retain the ordering. Finally, you can also pass a string representing your query string part. In this case, requests will prepend the question mark ("?") for you, but will — once again — not be able to provide smart URL encoding, so that you are responsible to make sure your query string is encoded properly. Although this is not frequently used, this can come in handy in cases where the web server expects an "?param" without an equals sign at the end, for instance — something that rarely occurs in practice, but can happen.

Silencing requests Completely   Even when passing a string to `params`, or including the full url in the `requests.get` method, requests will still try, as we have seen, to help out a little. For instance, writing:

```
requests.get('http://www.example.com/?spaces |pipe')
```

will make you end up with "?spaces%20%7Cpipe" as the query string in the request URL, with the space and pipe ("|") characters encoded for you. In rare situations, a very picky web server might nevertheless expect URLs to come in unencoded. Again, cases such as these are extremely rare, but we have encountered situations in the wild where this happens. In this case, you will need to override requests as follows:

```
import requests
from urllib.parse import unquote

class NonEncodedSession(requests.Session):
  # Override the default send method
  def send(self, *a, **kw):
    # Revert the encoding which was applied
    a[0].url = unquote(a[0].url)
    return requests.Session.send(self, *a, **kw)
```

```
my_requests = NonEncodedSession()
url = 'http://www.example.com/?spaces |pipe'
r = my_requests.get(url)
print(r.url)
# Will show: http://www.example.com/?spaces |pipe
```

As a final exercise, head over to *http://www.webscrapingfordatascience.com/ calchttp/*. Play around with the "a," "b," and "op" URL parameters. You should be able to work out what the following code does:

```
import requests

def calc(a, b, op):
    url = 'http://www.webscrapingfordatascience.com/calchttp/'
    params = {'a': a, 'b': b, 'op': op}
    r = requests.get(url, params=params)
    return r.text

print(calc(4, 6, '*'))
print(calc(4, 6, '/'))
```

Based on what we've seen above, you'll probably feel itchy to try out what you've learned using a real-life website. However, there is another hurdle we need to pass before being web ready. What happens, for instance, when you run the following:

```
import requests

url = 'https://en.wikipedia.org/w/index.php' + \
      '?title=List_of_Game_of_Thrones_episodes&oldid=802553687'
r = requests.get(url)
print(r.text)
```

Wikipedia Versioning   We're using the "oldid" URL parameter here such that we obtain a specific version of the "List of Game of Thrones episodes" page, to make sure that our subsequent examples will keep working. By the way, here you can see "URL rewriting" in action: both *https://en.wikipedia.org/wiki/List_of_Game_of_Thrones_episodes* and *https://en.wikipedia.org/w/index.php?title=List_of_Game_of_Thrones_episodes* lead to the exact same page. The difference is that the latter uses URL parameters and the former does not, though Wikipedia's web server is clever enough to route URLs to their proper "page." Also, you might note that we're not using the `params` argument here. We could, though neither the "title" nor "oldid" parameters require encoding here, so we can just stick them in the URL itself to keep the rest of the code a bit shorter.

As you can see, the response body captured by `r.text` now spits out a slew of confusing-looking text. This is HTML-formatted text, and although the content we're looking for is buried somewhere inside this soup, we'll need to learn about a proper way to get out the information we want from there. That's exactly what we'll do in the next chapter.

The Fragment Identifier   Apart from the query string, there is in fact another optional part of the URL that you might have encountered before: the fragment identifier, or "hash," as it is sometimes called. It is prepended by a hash mark ("#") and comes at the very end of a URL, even after the query string, for instance, as in "http://www.example.org/about.htm?p=8#contact". This part of the URL is meant to identify a portion of the document corresponding to the URL. For instance, a web page can include a link including a fragment identifier that, if you click on it, immediately scrolls your view to the corresponding part of the page. However, the fragment identifier functions differently than the rest of the URL, as it is processed exclusively by the web browser with no participation at all from the web server. In fact, proper web browsers should not even include the fragment identifier in their HTTP requests when they fetch a resource from a web server. Instead, the browser waits until the web server has sent its reply, and it will then use the fragment identifier to scroll to the correct part of the page. Most web servers will simply ignore a fragment identifier if you would include it in a request

47

URL, although some might be programmed to take them into account as well. Again: this is rather rare, as the content provided by such a server would not be viewable by most web browsers, as they leave out the fragment identifier part in their requests, though the web is full of interesting edge cases.

We've now seen the basics of the requests library. Take some time to explore the documentation of the library available at *http://docs.python-requests.org/en/master/*. The quality of requests' documentation is very high and easy to refer to once you start using the library in your projects.

CHAPTER 3

Stirring the HTML and CSS Soup

So far we have discussed the basics of HTTP and how you can perform HTTP requests in Python using the requests library. However, since most web pages are formatted using the Hypertext Markup Language (HTML), we need to understand how to extract information from such pages. As such, this chapter introduces you to HTML, as well as another core building block that is used to format and stylize modern web pages: Cascading Style Sheets (CSS). This chapter then discusses the Beautiful Soup library, which will help us to make sense of the HTML and CSS "soup."

3.1 Hypertext Markup Language: HTML

In the previous chapter, we introduced the basics of HTTP and saw how to perform HTTP requests in Python using the requests library, but now we need to figure out a way to parse HTML contents. Recall our small Wikipedia example we ended with in the previous chapter and the soup of HTML we got back from it:

```python
import requests

url = 'https://en.wikipedia.org/w/index.php' + \
      '?title=List_of_Game_of_Thrones_episodes&oldid=802553687'

r = requests.get(url)
print(r.text)
```

Perhaps you've tried running this example with some other favorite websites of yours... In any case, once you start looking a bit closer to how the web works and start web scraping in practice, you'll no doubt start to marvel at all the things your

© Seppe vanden Broucke and Bart Baesens 2018
S. vanden Broucke and B. Baesens, *Practical Web Scraping for Data Science*,
https://doi.org/10.1007/978-1-4842-3582-9_3

web browser does for you: getting out web pages; converting this "soup" into nicely formatted pages, including images, animation, styling, video, and so on. This might feel very intimidating at this point — surely we won't have to replicate all the things a web browser does from scratch? The answer is that, luckily, no we do not. Just as with HTTP, we'll use a powerful Python library that can help us navigate this textual mess. And, contrary to a web browser, we're not interested in fetching out a complete page's content and rendering it, but only in extracting those pieces we're interested in.

If you run the example above, you'll see the following being printed onscreen:

```
<!DOCTYPE html>
<html class="client-nojs" lang="en" dir="ltr">
<head>
<meta charset="UTF-8"/>
<title>List of Game of Thrones episodes - Wikipedia</title>
[...]
</html>
```

This is Hypertext Markup Language (HTML), the standard markup language for creating web pages. Although some will call HTML a "programming language," "markup language" is a more appropriate term as it specifies how a document is structured and formatted. There is no strict need to use HTML to format web pages — in fact, all the examples we've dealt with in the previous chapter just returned simple, textual pages. However, if you want to create visually appealing pages that actually look good in a browser (even if it's just putting some color on a page), HTML is the way to go.

HTML provides the building blocks to provide structure and formatting to documents. This is provided by means of a series of "tags." HTML tags often come in pairs and are enclosed in angled brackets, with "<tagname>" being the opening tag and "</tagname>" indicating the closing tag. Some tags come in an unpaired form, and do not require a closing tag. Some commonly used tags are the following:

- `<p>...</p>` to enclose a paragraph;

- `<br>` to set a line break;

- `<table>...</table>` to start a table block, inside; `<tr>...<tr/>` is used for the rows; and `<td>...</td>` cells;

- `<img>` for images;

- `<h1>...</h1>` to `<h6>...</h6>` for headers;

- `<div>...</div>` to indicate a "division" in an HTML document, basically used to group a set of elements;

- `<a>...</a>` for hyperlinks;

- `<ul>...</ul>`, `<ol>...</ol>` for unordered and ordered lists respectively; inside of these, `<li>...</li>` is used for each list item.

Tags can be nested inside each other, so "<div><p>Hello</p></div>" is perfectly valid, though overlapping nestings such as "<div><p>Oops</div></p>" is not. Even though this isn't proper HTML, every web browser will exert a lot of effort to still parse and render an HTML page as well as possible. If web browsers would require that all web pages are perfectly formatted according to the HTML standard, you can bet that the majority of websites would fail. HTML is messy.

Tags that come in pairs have content. For instance, "<a>click here</a>" will render out "click here" as a hyperlink in your browser. Tags can also have attributes, which are put inside of the opening tag. For instance, "<a href="`http://www.google.com`"> click here </a>" will redirect the user to Google's home page when the link is clicked. The "href" attribute hence indicates the web address of the link. For an image tag, which doesn't come in a pair, the "src" attribute is used to indicate the URL of the image the browser should retrieve, for example, "<img src="`http://www.example.com/image.jpg`">".

3.2 Using Your Browser as a Development Tool

Don't worry too much if all of this is going a bit fast, as we'll come to understand HTML in more detail when we work our way through the examples. Before we continue, we want to provide you with a few tips that will come in handy while building web scrapers.

Most modern web browsers nowadays include a toolkit of powerful tools you can use to get an idea of what's going on regarding HTML, and HTTP too. Navigate to the Wikipedia page over at *https://en.wikipedia.org/w/index.php?title=List_of_Game_of_Thrones_episodes&oldid=802553687* again in your browser — we assume you're using Google Chrome for what follows. First of all, it is helpful to know how you can take a look at the underlying HTML of this page. To do so, you can right-click on the page and press "View source," or simply press Control+U in Google Chrome. A new page will open containing the raw HTML contents for the current page (the same content as what we got back using `r.text`); see Figure 3-1.

Figure 3-1. *Viewing a page's source in Chrome*

Additionally, you can open up Chrome's "Developer Tools." To do so, either select the Chrome Menu at the top right of your browser window, then select "Tools," "Developer Tools," or press Control+Shift+I. Alternatively, you can also right-click on any page element and select "Inspect Element." Other browsers such as Firefox and Microsoft Edge have similar tools built in. You should get a screen like the one shown in Figure 3-2.

Figure 3-2. *The Chrome Developer Tools window contains a lot of helpful tools for web scrapers*

Moving Around   Take some time to explore the Developer Tools pane. Yours might appear at the bottom of your browser window. If you prefer to have it on the right, find the menu with the three-dotted-colon icon (the tri-colon), and pick a different "Dock side."

The Developer Tools pane is organized by means of a series of tabs, of which "Elements" and "Network" will come in most helpful.

Let's start by taking a look at the Network tab. You should see a red "recording" icon in the toolbar indicating that Chrome is tracking network requests (if the icon is not lit, press it to start tracking). Refresh the Wikipedia page and look at what happens in the Developer Tools pane: Chrome starts logging all requests it is making, starting with an HTTP request for the page itself at the top. Note that your web browser is also making lots of other requests to actually render the page, most of them to fetch image data ("Type: png"). By clicking a request, you can get more information about it. Click the "index.php" request at the top, for instance, to get a screen like in Figure 3-3. Selecting a request opens another pane that provides a wealth of information that should already look pretty familiar to you now that you've already worked with HTTP. For instance, making sure the "Headers" tab is selected in the side pane, we see general information such as the request URL, method (verb), and status code that was sent back by the server, as well as a full list of request and response headers.

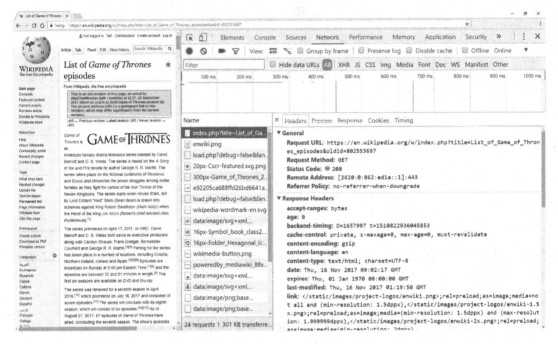

Figure 3-3. *Inspecting an HTTP request in Chrome*

Finally, there are a number of useful check boxes in the Network tab that are noteworthy to mention. Enabling "Preserve log" will prevent Chrome from "cleaning up" the overview every time a new page request is performed. This can come in handy in case you want to track a series of actions when navigating a website. "Disable cache" will prevent Chrome from using its "short-term memory." Chrome will try to be smart and prevent performing a request if it still has the contents of a recent page around, though you can override this in case you want to force Chrome to actually perform every request.

Moving on to the "Elements" tab, we see a similar view as what we see when viewing the page's source, though now neatly formatted as a tree-based view, with little arrows that we can expand and collapse, as shown in Figure 3-4.

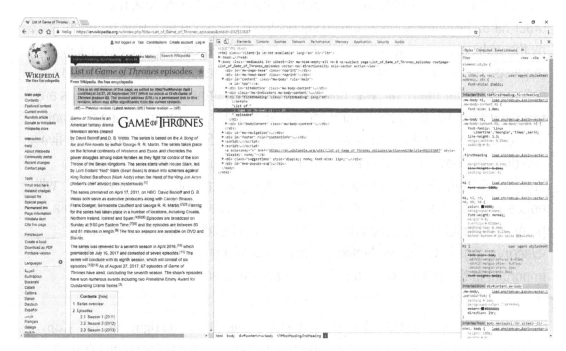

Figure 3-4. *Looking at the Elements tab in Chrome*

What is particularly helpful here is the fact that you can hover over the HTML tags in the Elements tab, and Chrome will show a transparent box over the corresponding visual representation on the web page itself. This can help you to quickly find the pieces of content you're looking for. Alternatively, you can right-click any element on a web page and press "Inspect element" to immediately highlight its corresponding HTML code in the Elements tab. Note that the "breadcrumb trail" at the bottom of the Elements tab shows you where you currently are in the HTML "tree."

Inspecting Elements versus View Source   You might wonder why the "View source" option is useful to look at a page's raw HTML source when we have a much user-friendlier alternative offered by the Elements tab. A warning is in order here: the "View source" option shows the HTML code as it was returned by the web server, and it will contain the same contents as `r.text` when using requests. The view in the Elements tab, on the other hand, provides a "cleaned up" version after the HTML was parsed by your web browser. Overlapping tags are fixed and extra white space is removed, for instance. There might hence be small differences between these two views. In addition, the Elements tab provides a live and dynamic view. Websites can include scripts that are executed by your web browser and which can alter the contents of the page at will. The Elements tab will hence always reflect the current state of the page. These scripts are written in a programming language called JavaScript and can be found inside <script>... </script> tags in HTML. We'll take a closer look at JavaScript and why it is important in the context of web scraping a few chapters later.

Next, note that any HTML element in the Elements tab can be right-clicked. "Copy, Copy selector" and "Copy XPath" are particularly useful, which we're going to use quite often later on. You'll even see that you can edit the HTML code in real time (the web page will update itself to reflect your edits), though don't feel too much like a *CSI: Miami* style hacker: these changes are of course only local. They don't do anything on the web server itself and will be gone once you refresh the page, though it can be a fun way to experiment with HTML. In any case, your web browser is going to become your best friend when working on web scraping projects.

3.3 Cascading Style Sheets: CSS

Before we can get started with actually dealing with HTML in Python, there is another key piece of technology that we need to discuss first: Cascading Style Sheets (CSS). While perusing the HTML elements in your browser, you've probably noticed that some HTML attributes are present in lots of tags:

- "id," which is used to attach a page-unique identifier to a certain tag;
- "class," which lists a space-separated series of CSS class names.

Whereas "id" will come in handy to quickly fetch parts of an HTML page we're interested in, "class" deserves some further introduction and relates to the concept of CSS.

CSS and HTML go hand in hand. Recall that, originally, HTML was meant as a way to define both the structure and formatting of a website. In the early days of the web, it was hence normal to find lots of HTML tags that were meant to define what content should look like, for example "..." for bold text; "<i>...</i>" for italics text; and "<font>...</font>" to change the font family, size, color, and other font attributes. After a while, however, web developers began to argue — rightly so — that the structure and formatting of documents basically relate to two different concerns. Compare this to writing a document with a text processor such as Word. You can either apply formatting directly to the document, but a better way to approach this is to use styles to indicate headers, lists, tables, etc., the formatting of which can then easily be changed by modifying the definition of the style. CSS works in a similar way. HTML is still used to define the general structure and semantics of a document, whereas CSS will govern how a document should be styled, or in other words, what it should look like.

The CSS language looks somewhat different from HTML. In CSS, style information is written down as a list of colon-separated key-value based statements, with each statement itself being separated by a semicolon, as follows:

```
color: 'red';
background-color: #ccc;
font-size: 14pt;
border: 2px solid yellow;
```

These style declarations can be included in a document in three different ways:

- Inside a regular HTML "style" attribute, for instance as in: "<p style="color:'red';">...</p>".

- Inside of HTML "<style>...</style>" tags, placed in inside the "<head>" tag of a page.

- Inside a separate file, which is then referred to by means of a "<link>" tag inside the "<head>" tag of a page. This is the cleanest way of working. When loading a web page, your browser will perform an additional HTTP request to download this CSS file and apply its defined styles to the document.

In case style declarations are placed inside a "style" attribute, it is clear to which element the declarations should be applied: the HTML tag itself. In the other two cases, the style definition needs to incorporate information regarding the HTML element or elements a styling should be applied to. This is done by placing the style declarations inside curly brackets to group them, and putting a "CSS selector" at the beginning of each group:

```
h1 {
  color: red;
}
div.box {
  border: 1px solid black;
}
#intro-paragraph {
  font-weight: bold;
}
```

CSS selectors define the patterns used to "select" the HTML elements you want to style. They are quite comprehensive in terms of syntax. The following list provides a full reference:

- tagname selects all elements with a particular tag name. For instance, "h1" simply matches with all "<h1>" tags on a page.

- .classname (note the dot) selects all elements having a particular class defined in the HTML document. This is exactly where the "class" attribute comes in. For instance, .intro will match with both "<p class="intro">" and "<h1 class="intro">". Note that HTML elements can have multiple classes, for example, "<p class="intro highlight">".

- #idname matches elements based on their "id" attribute. Contrary to classes, proper HTML documents should ensure that each "id" is unique and only given to one element only (though don't be surprised if some particularly messy HTML page breaks this convention and used the same id value multiple times).

- These selectors can be combined in all sorts of ways. div.box, for instance, selects all "<div class="box">" tags, but not "<div class="circle">" tags.

- Multiple selector rules can be specified by using a comma, ",", for example, `h1, h2, h3`.

- `selector1 selector2` defines a chaining rule (note the space) and selects all elements matching `selector2` inside of elements matching `selector1`. Note that it is possible to chain more than two selectors together.

- `selector1 > selector2` selects all elements matching `selector2` where the parent element matches `selector1`. Note the subtle difference here with the previous line. A "parent" element refers to the "direct parent." For instance, `div > span` will not match with the span element inside "<div> <p> <span> </span> </p> </div>" (as the parent element here is a "<p>" tag), whereas `div span` will.

- `selector1 + selector2` selects all elements matching `selector2` that are placed directly after (i.e., on the same level in the HTML hierarchy) elements matching `selector1`.

- `selector1 ~ selector2` selects all elements matching `selector2` that are placed after (on the same level in the HTML hierarchy) `selector1`. Again, there's a subtle difference here with the previous rule: the precedence here does not need to be "direct": there can be other tags in between.

- It is also possible to add more fine-tuned selection rules based on attributes of elements. `tagname[attributename]` selects all `tagname` elements where an attribute named `attributename` is present. Note that the tag selector is optional, and simply writing `[title]` selects all elements with a "title" attribute.

- The attribute selector can be further refined. `[attributename=value]` checks the actual value of an attribute as well. If you want to include spaces, wrap the value in double quotes.

- `[attributename~=value]` does something similar, but instead of performing an exact value comparison, here all elements are selected whose `attributename` attribute's value is a space-separated list of words, one of them being equal to `value`.

- `[attributename|=value]` selects all elements whose `attributename` attribute's value is a space-separated list of words, with any of them being equal to "value" or starting with "value" and followed by a hypen ("-").

- `[attributename^=value]` selects all elements whose attribute value starts with the provided value. If you want to include spaces, wrap the value in double quotes.

- `[attributename$=value]` selects all elements whose attribute value ends with the provided value. If you want to include spaces, wrap the value in double quotes.

- `[attributename*=value]` selects all elements whose attribute value contains the provided value. If you want to include spaces, wrap the value in double quotes.

- Finally, there are a number of "colon" and "double-colon" "pseudo-classes" that can be used in a selector rule as well. `p:first-child` selects every "<p>" tag that is the first child of its parent element, and `p:last-child` and `p:nth-child(10)` provide similar functionality.

Play around with the Wikipedia page using your Chrome's Developer Tools (or the equivalent in your browser): try to find instances of the "class" attribute. The CSS resource of the page is referenced through a "<link>" tag (note that pages can load multiple CSS files as well):

```
<link rel="stylesheet" href="/w/load.php?[...];skin=vector">
```

We're not going to build websites using CSS. Instead, we're going to scrape them. As such, you might wonder why this discussion regarding CSS is useful for our purposes. The reason is that the same CSS selector syntax can be used to quickly find and retrieve elements from an HTML page using Python. Try right-clicking some HTML elements in the Elements tab of Chrome's Developer Tools pane and press "Copy, Copy selector." Note that you obtain a CSS selector. For instance, this is the selector to fetch one of the tables on the page:

```
#mw-content-text > div > table:nth-child(9).
```

Or: "inside the element with id "mw-content-text," get the child "div" element, and get the 9th "table" child element." We'll use these selectors quite often once we start working with HTML in our web scraping scripts.

3.4 The Beautiful Soup Library

We're now ready to start working with HTML pages using Python. Recall the following lines of code:

```python
import requests

url = 'https://en.wikipedia.org/w/index.php' + \
      '?title=List_of_Game_of_Thrones_episodes&oldid=802553687'

r = requests.get(url)
html_contents = r.text
```

How do we deal with the HTML contained in html_contents? To properly parse and tackle this "soup," we'll bring in another library, called "Beautiful Soup."

Soup, Rich and Green   And finally, it becomes clear why we've been referring to messy HTML pages as a "soup." The Beautiful Soup library was named after a Lewis Carroll poem bearing the same name from "Alice's Adventures in Wonderland." In the tale, the poem is sung by a character called the "Mock Turtle" and goes as follows: "Beautiful Soup, so rich and green,// Waiting in a hot tureen!// Who for such dainties would not stoop?// Soup of the evening, beautiful Soup!". Just like in the story, Beautiful Soup tries to organize complexity: it helps to parse, structure and organize the oftentimes very messy web by fixing bad HTML and presenting us with an easy-to-work-with Python structure.

Just as was the case with requests, installing Beautiful Soup is easy with pip (refer back to section 1.2.1 if you still need to set up Python 3 and pip), and note the "4" in the package name:

```
pip install -U beautifulsoup4
```

Using Beautiful Soup starts with the creation of a `BeautifulSoup` object. If you already have an HTML page contained in a string (as we have), this is straightforward. Don't forget to add the new import line:

```
import requests
from bs4 import BeautifulSoup

url = 'https://en.wikipedia.org/w/index.php' + \
      '?title=List_of_Game_of_Thrones_episodes&oldid=802553687'

r = requests.get(url)
html_contents = r.text

html_soup = BeautifulSoup(html_contents)
```

Try running this snippet of code. If everything went well, you should get no errors, though you might see the following warning appear:

```
Warning (from warnings module):
  File "__init__.py", line 181      markup_type=markup_type))

UserWarning: No parser was explicitly specified, so I'm using the best ⏎
available HTML parser for this system ("html.parser"). This usually ⏎
isn't a problem, but if you run this code on another system, or in a ⏎
different virtual environment, it may use a different parser and behave ⏎
differently.

The code that caused this warning is on line 1 of the file <string>. ⏎
To get rid of this warning, change code so that it looks like this:

 BeautifulSoup(YOUR_MARKUP})

to this:

 BeautifulSoup(YOUR_MARKUP, "html.parser")
```

Uh-oh, what's going on here? The Beautiful Soup library itself depends on an HTML parser to perform most of the bulk parsing work. In Python, multiple parsers exist to do so:

- "html.parser": a built-in Python parser that is decent (especially when using recent versions of Python 3) and requires no extra installation.

- "lxml": which is very fast but requires an extra installation.

- "html5lib": which aims to parse web page in exactly the same way as a web browser does, but is a bit slower.

Since there are small differences between these parsers, Beautiful Soup warns you if you don't explicitly provide one, this might cause your code to behave slightly different when executing the same script on different machines. To solve this, we simply specify a parser ourselves — we'll stick with the default Python parser here:

```
html_soup = BeautifulSoup(html_contents, 'html.parser')
```

Beautiful Soup's main task is to take HTML content and transform it into a tree-based representation. Once you've created a `BeautifulSoup` object, there are two methods you'll be using to fetch data from the page:

- `find(name, attrs, recursive, string, **keywords);`

- `find_all(name, attrs, recursive, string, limit, **keywords).`

Underscores   If you don't like writing underscores, Beautiful Soup also exposes most of its methods using "camelCaps" capitalization. So instead of `find_all`, you can also use `findAll` if you prefer.

Both methods look very similar indeed, with the exception that `find_all` takes an extra `limit` argument. To test these methods, add the following lines to your script and run it:

```
print(html_soup.find('h1'))

print(html_soup.find('', {'id': 'p-logo'}))

for found in html_soup.find_all(['h1', 'h2']):
    print(found)
```

The general idea behind these two methods should be relatively clear: they're used to find elements inside the HTML tree. Let's discuss the arguments of these two methods step by step:

- The name argument defines the tag names you wish to "find" on the page. You can pass a string, or a list of tags. Leaving this argument as an empty string simply selects all elements.

- The attrs argument takes a Python dictionary of attributes and matches HTML elements that match those attributes.

And or Or?   Some guidelines state that the attributes defined in the attrs dictionary behave in an "or-this-or-that" relationship, where every element that matches at least one of the attributes will be retrieved. This is not true, however: both your filters defined in attrs and in the keywords you use in **keywords should all match in order for an element to be retrieved.

- The recursive argument is a Boolean and governs the depth of the search. If set to True — the default value, the find and find_all methods will look into children, children's children, and so on... for elements that match your query. If it is False, it will only look at direct child elements.

- The string argument is used to perform matching based on the text content of elements.

Text or String?   The string argument is relatively new. In earlier Beautiful Soup versions, this argument was named text instead. You can, in fact, still use text instead of string if you like. If you use both (not recommended), then text takes precedence and string ends up in the list of **keywords below.

- The limit argument is only used in the find_all method and can be used to limit the number of elements that are retrieved. Note that find is functionally equivalent to calling find_all with the limit set to 1, with the exception that the former returns the retrieved element directly, and that the latter will always return a list of items, even if it just contains a single element. Also important to know is that, when find_all cannot find anything, it returns an empty list, whereas if find cannot find anything, it returns None.

- **keywords is kind of a special case. Basically, this part of the method signature indicates that you can add in as many extra named arguments as you like, which will then simply be used as attribute filters. Writing "find(id='myid')" is hence the same as "find(attrs={'id': 'myid'})". If you define both the attrs argument and extra keywords, all of these will be used together as filters. This functionality is mainly offered as a convenience in order to write easier-to-read code.

Take Care with Keywords   Even although the **keywords argument can come in very helpful in practice, there are some important caveats to mention here. First of all, you cannot use class as a keyword, as this is a reserved Python keyword. This is a pity, as this will be one of the most frequently used attributes when hunting for content inside HTML. Luckily, Beautiful Soup has provided a workaround. Instead of using class, just write class_ as follows: "find(class_='myclass')". Note that name can also not be used as a keyword, since that is what is used already as the first argument name for find and find_all. Sadly, Beautiful Soup does not provide a name_ alternative here. Instead, you'll need to use attrs in case you want to select based on the "name" HTML attribute.

Both find and find_all return Tag objects. Using these, there are a number of interesting things you can do:

- Access the name attribute to retrieve the tag name.

- Access the contents attribute to get a Python list containing the tag's children (its direct descendant tags) as a list.

- The children attribute does the same but provides an iterator instead; the descendants attribute also returns an iterator, now including all the tag's descendants in a recursive manner. These attributes are used when you call find and find_all.

- Similarly, you can also go "up" the HTML tree by using the parent and parents attributes. To go sideways (i.e., find next and previous elements at the same level in the hierarchy), next_sibling, previous_sibling and next_siblings, and previous_siblings can be used.

- Converting the Tag object to a string shows both the tag and its HTML content as a string. This is what happens if you call print out the Tag object, for instance, or wrap such an object in the str function.

- Access the attributes of the element through the attrs attribute of the Tag object. For the sake of convenience, you can also directly use the Tag object itself as a dictionary.

- Use the text attribute to get the contents of the Tag object as clear text (without HTML tags).

- Alternatively, you can use the get_text method as well, to which a strip Boolean argument can be given so that get_text(strip=True) is equivalent to text.strip(). It's also possible to specify a string to be used to join the bits of text enclosed in the element together, for example, get_text('--').

- If a tag only has one child, and that child itself is simply text, then you can also use the string attribute to get the textual content. However, in case a tag contains other HTML tags nested within, string will return None whereas text will recursively fetch all the text.

- Finally, not all find and find_all searches need to start from your original BeautifulSoup objects. Every Tag object itself can be used as a new root from which new searches can be started.

We've dealt with a lot of theory. Let's show off these concepts through some example code:

```python
import requests
from bs4 import BeautifulSoup

url = 'https://en.wikipedia.org/w/index.php' + \
      '?title=List_of_Game_of_Thrones_episodes&oldid=802553687'

r = requests.get(url)
html_contents = r.text
html_soup = BeautifulSoup(html_contents, 'html.parser')
# Find the first h1 tag
first_h1 = html_soup.find('h1')

print(first_h1.name)      # h1
print(first_h1.contents) # ['List of ', [...], ' episodes']

print(str(first_h1))
# Prints out: <h1 class="firstHeading" id="firstHeading" lang="en">List of
#             <i>Game of Thrones</i> episodes</h1>

print(first_h1.text)       # List of Game of Thrones episodes
print(first_h1.get_text()) # Does the same

print(first_h1.attrs)
# Prints out: {'id': 'firstHeading', 'class': ['firstHeading'], 'lang': 'en'}

print(first_h1.attrs['id']) # firstHeading
print(first_h1['id'])        # Does the same
print(first_h1.get('id'))    # Does the same

print('------------ CITATIONS ------------')
# Find the first five cite elements with a citation class
cites = html_soup.find_all('cite', class_='citation', limit=5)
for citation in cites:
    print(citation.get_text())
    # Inside of this cite element, find the first a tag
    link = citation.find('a')
```

```
# ... and show its URL
print(link.get('href'))
print()
```

A Note About Robustness   Take a good look at the "citations" part of the example above. What would happen in case no "<a>" tag is present inside a "<cite>" element? In that case, the link variable would be set to None and the line "link.get('href')" would crash our program. Always take care when writing web scrapers and prepare for the worst. For examples in "safe environments" we can permit ourselves to be somewhat sloppy for the sake of brevity, but in a real-life situation, you'd want to put in an extra check to see whether link is None or not and act accordingly.

Before we move on with another example, there are two small remarks left to be made regarding find and find_all. If you find yourself traversing a chain of tag names as follows:

```
tag.find('div').find('table').find('thead').find('tr')
```

It might be useful to keep in mind that Beautiful Soup also allows us to write this in a shorthand way:

```
tag.div.table.thead.tr
```

Similarly, the following line of code:

```
tag.find_all('h1')
```

Is the same as calling:

```
tag('h1')
```

Although this is — again — offered for the sake of convenience, we'll nevertheless continue to use find and find_all in full throughout this book, as we find that being a little bit more explicit helps readability in this case.

Let us now try to work out the following use case. You'll note that our *Game of Thrones* Wikipedia page has a number of well-maintained tables listing the episodes with their directors, writers, air date, and number of viewers. Let's try to fetch all of this data at once using what we have learned:

```python
import requests
from bs4 import BeautifulSoup

url = 'https://en.wikipedia.org/w/index.php' + \
      '?title=List_of_Game_of_Thrones_episodes&oldid=802553687'

r = requests.get(url)
html_contents = r.text
html_soup = BeautifulSoup(html_contents, 'html.parser')

# We'll use a list to store our episode list
episodes = []

ep_tables = html_soup.find_all('table', class_='wikiepisodetable')

for table in ep_tables:
    headers = []
    rows = table.find_all('tr')
    # Start by fetching the header cells from the first row to determine
    # the field names
    for header in table.find('tr').find_all('th'):
        headers.append(header.text)
    # Then go through all the rows except the first one
    for row in table.find_all('tr')[1:]:
        values = []
        # And get the column cells, the first one being inside a th-tag
        for col in row.find_all(['th','td']):
            values.append(col.text)
        if values:
            episode_dict = {headers[i]: values[i] for i in
            range(len(values))}
            episodes.append(episode_dict)

# Show the results
for episode in episodes:
    print(episode)
```

Most of the code should be relatively straightforward at this point, though some things are worth pointing out:

- We don't come up with the "find_all('table', class_ = 'wikiepisodetable')" line from thin air, although it might seem that way just by looking at the code. Recall what we said earlier about your browser's developer tools becoming your best friend. Inspect the episode tables on the page. Note how they're all defined by means of a "<table>" tag. However, the page also contains tables we do not want to include. Some further investigation leads us to a solution: all the episode tables have "wikiepisodetable" as a class, whereas the other tables do not. You'll often have to puzzle your way through a page first before coming up with a solid approach. In many cases, you'll have to perform multiple find and find_all iterations before ending up where you want to be.

- For every table, we first want to retrieve the headers to use as keys in a Python dictionary. To do so, we first select the first "<tr>" tag, and select all "<th>" tags within it.

- Next, we loop through all the rows (the "<tr>" tags), except for the first one (the header row). For each row, we loop through the "<th>" and "<td>" tags to extract the column values (the first column is wrapped inside of a "<th>" tag, the others in "<td>" tags, which is why we need to handle both). At the end of each row, we're ready to add a new entry to the "episodes" variable. To store each entry, we use a normal Python dictionary (episode_dict). The way how this object is constructed might look a bit strange in case you're not very familiar with Python. That is, Python allows us to construct a complete list or dictionary "in one go" by putting a "for" construct inside the "[...]" or "{...}" brackets. Here, we use this to immediately loop through the headers and values lists to build the dictionary object. Note that this assumes that both of these lists have the same length, and that the order for both of these matches so that the header at "headers[2]", for instance, is the header corresponding with the value over at "values[2]". Since we're dealing with rather simple tables here, this is a safe assumption.

Are Tables Worth It?   You might not be very impressed with this example so far. Most modern browsers allow you to simply select or right-click tables on web pages and will be able to copy them straight into a spreadsheet program such as Excel anyway. That's true, and if you only have one table to extract, this is definitely the easier route to follow. Once you start dealing with many tables, however, especially if they're spread over multiple pages, or need to periodically refresh tabular data from a particular web page, the benefit of writing a scraper starts to become more apparent.

Experiment a bit more with this code snippet. You should be able to work out the following:

- Try extracting all links from the page as well as where they point to (tip: look for the "href" attribute in "<a>" tags).

- Try extracting all images from the page.

- Try extracting the "ratings" table from the page. This one is a bit tricky. You might be tempted to use "find('table', class_="wikitable")", but you'll note that this matches the very first table on the page instead, even though its class attribute is set to "wikitable plainrowheaders." Indeed, for HTML attributes that can take multiple, space-separated values (such as "class"), Beautiful Soup will perform a partial match. To get out the table we want, you'll either have to loop over all "wikitable" tables on the page and perform a check on its text attribute to make sure you have the one you want, or try to find a unique parent element from which you can drill down, for example, "find('div', align="center"). find('table', class_="wikitable")" — at least for now, you'll learn about some more advanced Beautiful Soup features in the upcoming section.

Classes Are Special   For HTML attributes that can take multiple, space-separated values (such as "class"), Beautiful Soup will perform a partial match. This can be tricky in case you want to perform an exact match such as "find me elements with the "myclass" class and only that class," but also in cases where you want to select an HTML element matching more than one class. In this case, you can write something like "find(class_='class-one class-two')", though this way of working is rather brittle and dangerous (the classes should then appear in the same order and next to each other in the HTML page, which might not always be the case). Another approach is to wrap your filter in a list, that is, "find(class_=['class-one', 'class-two'])", though this will also not obtain the desired result: instead of matching elements having both "class-one" and "class-two" as classes, this statement will match with elements having any of these classes! To solve this problem in a robust way, we hence first need to learn a little bit more about Beautiful Soup...

3.5 More on Beautiful Soup

Now that we understand the basics of Beautiful Soup, we're ready to explore the library a bit further. First of all, although we have already seen the basics of find and find_all, it is important to note that these methods are very versatile indeed. We have already seen how you can filter on a simple tag name or a list of them:

```
html_soup.find('h1')
html_soup.find(['h1', 'h2'])
```

However, these methods can also take other kinds of objects, such as a regular expression object. The following line of code will match with all tags that start with the letter "h" by constructing a regular expression using Python's "re" module:

```
import re

html_soup.find(re.compile('^h'))
```

Regex   If you haven't heard about regular expressions before, a regular expression (regex) defines a sequence of patterns (an expression) defining a search pattern. It is frequently used for string searching and matching code to find (and replace) fragments of strings. Although they are very powerful constructs, they can also be misused. For instance, it is a good idea not to go overboard with long or complex regular expressions, as they're not very readable and it might be hard to figure out what a particular piece of regex is doing later on. By the way, this is also a good point to mention that you should avoid using regex to parse HTML pages. We could have introduced regex earlier as a way to parse pieces of content from our HTML soup, without resorting at all to the use of Beautiful Soup. This is a terrible idea, however. HTML pages are — as we have seen — very messy, and you'll quickly end up with lots of spaghetti code in order to extract content from a page. Always use an HTML parser like Beautiful Soup to perform the grunt work. You can then use small snippets of regex (as shown here) to find or extract pieces of content.

Apart from strings, lists, and regular expressions, you can also pass a function. This is helpful in complicated cases where other approaches wouldn't work:

```python
def has_classa_but_not_classb(tag):
    cls = tag.get('class', [])
    return 'classa' in cls and not 'classb' in cls

html_soup.find(has_classa_but_not_classb)
```

Note that you can also pass lists, regular expressions, and functions to the attrs dictionary values, string, and **keyword arguments of find and find_all.

Apart from find and find_all, there are also a number of other methods for searching the HTML tree, which are very similar to find and find_all. The difference is that they will search different parts of the HTML tree:

- find_parent and find_parents work their way up the tree, looking at a tag's parents using its parents attribute. Remember that find and find_all work their way down the tree, looking at a tag's descendants.

- find_next_sibling and find_next_siblings will iterate and match a tag's siblings using next_siblings attribute.

- find_previous_sibling and find_previous_siblings do the same but use the previous_siblings attribute.

- find_next and find_all_next use the next_elements attribute to iterate and match over whatever comes after a tag in the document.

- find_previous and find_all_previous will perform the search backward using the previous_elements attribute instead.

- Remember that find and find_all work on the children attribute in case the recursive argument is set to False, and on the descendants attribute in case recursive is set to True.

So Many Methods   Although it's not really documented, it is also possible to use the findChild and findChildren methods (though not find_child and find_children), which are defined as aliases for find and find_all respectively. There is no findDescendant, however, so keep in mind that using findChild will default to searching throughout all the descendants (just like find does), unless you set the recursive argument to False. This is certainly confusing, so it's best to avoid these methods.

Although all of these can come in handy, you'll find that find and find_all will take up most of the workload when navigating an HTML tree. There is one more method, however, which is extremely useful: select. Finally, the CSS selectors we've seen above can be of use. Using this method, you can simply pass a CSS selector rule as a string.

Beautiful Soup will return a list of elements matching this rule:

```
# Find all <a> tags
html_soup.select('a')

# Find the element with the info id
html_soup.select('#info')

# Find <div> tags with both classa and classb CSS classes
html_soup.select(div.classa.classb)
```

```
# Find <a> tags with an href attribute starting with http://example.com/
html_soup.select('a[href^="http://example.com/"]')
```

```
# Find <li> tags which are children of <ul> tags with class lst
html_soup.select(ul.lst > li')
```

Once you start getting used to CSS selectors, this method can be very powerful indeed. For instance, if we want to find out the citation links from our *Game of Thrones* Wikipedia page, we can simply run:

```
for link in html_soup.select('ol.references cite a[href]'):
    print(link.get('href'))
```

However, the CSS selector rule engine in Beautiful Soup is not as powerful as the one found in a modern web browser. The following rules are valid selectors, but will not work in Beautiful Soup:

```
# This will not work:
# cite a[href][rel=nofollow]
```

```
# Instead, you can use:
tags = [t for t in html_soup.select('cite a[href]') \
        if 'nofollow' in t.get('rel', [])]
```

```
# This will not work:
# cite a[href][rel=nofollow]:not([href*="archive.org"])
```

```
# Instead, you can use:
tags = [t for t in html_soup.select('cite a[href]') \
        if 'nofollow' in t.get('rel', []) and 'archive.org' not in
            t.get('href')]
```

Luckily, cases where you need to resort to such complex selectors are rare, and remember that you can still use find, find_all, and friends too (try playing around with the two samples above and rewrite them without using select at all).

All Roads Lead to Your Element   Observant readers will have noticed that there are often multiple ways to write a CSS selector to get the same result. That is, instead of writing "cite a," you can also go overboard and write "body div.reflist ol.references li cite.citation a" and get the same result. In general, however, it is good practice to only make your selectors as granular as necessary to get the content you want. Websites often change, and if you're planning to use a web scraper in production for a significant amount of time, you can save yourself some headache by finding a good trade-off between precision and robustness. That way, you can try to be as future proof as possible in case the site owners decide to play around with the HTML structure, class names, attributes, and so on. This being said, there might always be a moment where a change is so significant that it ends up breaking your selectors. Including extra checks in your code and providing early warning signs can help a lot to build robust web scrapers.

Finally, there is one more detail to mention regarding Beautiful Soup. So far, we've been talking mostly about the BeautifulSoup object itself (the html_soup variable in our examples above), as well as Tag objects retrieved by find, find_all, and other search operations. There are two more object types in Beautiful Soup that, although less commonly used, are useful to know about:

- NavigableString objects: these are used to represent text within tags, rather than the tags themselves. Some Beautiful Soup functions and attributes will return such objects, such as the string attribute of tags, for instance. Attributes such as descendants will also include these in their listings. In addition, if you use find or find_all methods and supply a string argument value without a name argument, then these will return NavigableString objects as well, instead of Tag objects.

- Comment objects: these are used to represent HTML comments (found in comment tags, "<!-- ... -->"). These are very rarely useful when web scraping.

Feel free to play around some more with Beautiful Soup if you like. Take some time to explore the documentation of the library over at *https://www.crummy.com/software/ BeautifulSoup/bs4/doc/*. Note that Beautiful Soup's documentation is a bit less well-structured than requests' documentation, so it reads more like an end-to-end document instead of a reference guide. In the next chapter, we take a step back from HTML and return to HTTP to explore it in more depth.

PART II

Advanced Web Scraping

PART II

Advanced Web Scraping

CHAPTER 4

Delving Deeper in HTTP

We've already seen most of the core building blocks that make up the modern web: HTTP, HTML, and CSS. However, we're not completely finished with HTTP yet. So far, we've only been using one of HTTP's request "verbs" or "methods": "GET". This chapter will introduce you to the other methods HTTP provides, starting with the "POST" method that is commonly used to submit web forms. Next, this chapter explores HTTP request and reply headers in more depth, and shows how you can deal with cookies using the requests library. The chapter closes with a discussion on other common forms of content other than HTML formatted pages you will frequently encounter on the web, and how to deal with them in your web scraping projects.

4.1 Working with Forms and POST Requests

We've already seen one way how web browsers (and you) can pass input to a web server, that is, by simply including it in the requested URL itself, either by including URL parameters or simply by means of the URL path itself, as was previously discussed. However, one can easily argue that this way of providing input is not that user friendly. Imagine that we'd want to buy some tickets for a concert, and that we'd be asked to send our details to a web server by including our name, e-mail address, and other information as a bunch of URL parameters. Not a very pleasant idea indeed! In addition, URLs are (by definition) limited in terms of length, so in case we want to send lots of information to a web server, this "solution" would also fail to work.

Websites provide a much better way to facilitate providing input and sending that input to a web server, one that you have no doubt already encountered: web forms. Whether it is to provide a "newsletter sign up" form, a "buy ticket" form, or simply a "login" form, web forms are used to collect the appropriate data. The way how web

© Seppe vanden Broucke and Bart Baesens 2018
S. vanden Broucke and B. Baesens, *Practical Web Scraping for Data Science*,
https://doi.org/10.1007/978-1-4842-3582-9_4

forms are shown in a web browser is simply by including the appropriate tags inside of HTML. That is, each web form on a page corresponds with a block of HTML code enclosed in "<form>" tags:

```
<form>
[...]
</form>
```

Inside of these, there are a number of tags that represent the form fields themselves. Most of these are provided through an "<input>" tag, with the "type" attribute specifying what kind of field it should represent:

- `<input type="text">` for simple text fields;

- `<input type="password">` for password entry fields;

- `<input type="button">` for general-purpose buttons;

- `<input type="reset">` for a "reset" button (when clicked, the browser will reset all form values to their initial state, but this button is rarely encountered these days);

- `<input type="submit">` for a "submit" button (more on this later);

- `<input type="checkbox">` for check boxes;

- `<input type="radio">` for radio boxes;

- `<input type="hidden">` for hidden fields, which will not be shown to the user but can still contain a value.

Apart from these, you'll also find pairs of tags being used ("<input>" does not come with a closing tag):

- `<button>...</button>` as another way to define buttons;

- `<select>...</select>` for drop-down lists. Within these, every choice is defined by using `<option>...</option>` tags;

- `<textarea>...</textarea>` for larger text entry fields.

Navigate to *http://www.webscrapingfordatascience.com/basicform/* to see a basic web form in action (see Figure 4-1).

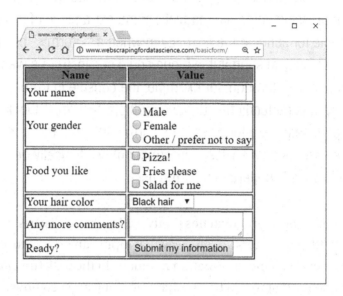

Figure 4-1. *A simple web form illustrating the different input fields*

Take some time to inspect the corresponding HTML source using your web browser. You'll notice that some HTML attributes seem to play a particular role here, that is, the "name" and "value" attributes for the form tags. As it turns out, these attributes will be used by your web browser once a web form is "submitted." To do so, try pressing the "Submit my information" button on the example web page (feel free to fill in some dummy information first). What do you see? Notice that upon submitting a form, your browser fires of a new HTTP request and includes the entered information in its request. In this simple form, a simple HTTP GET request is being used, basically converting the fields in the form to key-value URL parameters.

Submitting Forms   A "submit" button is not the only way that form information can be submitted. Sometimes, forms consist of simply one text field, which will be submitted when pressing enter. Alternatively, some extra piece of JavaScript can also be responsible for sending the form. Even when a "<form>" block does not expose any obvious way to submit itself, one could still instruct a web browser to submit the form anyway, through, for example, the console in Chrome's Developer Tools. By the way: it is perfectly fine to have multiple "<form>" blocks inside of the same web page (e.g., one for a "search" form and one for a "contact us" form), though only one form (and its enclosed information) will typically be submitted once a "submit" action is undertaken by the user.

This way of submitting a form matches pretty well with our discussion from a few paragraphs back: URL parameters are one way that input can be sent to a web server. Instead of having to enter all the information manually in the URL (imagine what a horrible user experience that would be), we have now at least seen how web forms can make this process a little bit user friendlier.

However, in cases where we have to submit a lot of information (try filling the "comments" form field on the example page with lots of text, for instance), URLs become unusable to submit information, due to their maximum length restriction. Even if URLs would be unbounded in terms of length, they would still not cater to a fully appropriate mechanism to submit information. What would happen, for instance, if you copy-paste such a URL in an e-mail and someone else would click on it, hence "sending" the information once again to the server? Or what happens in case you accidentally refresh such an URL? In a case where we are sending information to a web server of which we can expect that this submission action will make a permanent change (such as deleting a user, changing your profile information, clicking "accept" for a bank transfer, and so on), it is perhaps not a good idea to simply allow for such requests to come in as HTTP GET requests.

Luckily, the HTTP protocol also provides a number of different "methods" (or "verbs") other than the GET method we've been working with so far. More specifically, apart from GET requests, there is another type of request that your browser will often be using in case you wish to submit some information to a web server: the POST request. To introduce it, navigate to the page over at *http://www.webscrapingfordatascience.com/postform/*. Notice that this page looks exactly the same as the one from above, except for one small difference in the HTML source: there's an extra "method" attribute for the "<form>" tag now:

```
<form method="post">
[...]
</form>
```

The default value for the "method" attribute is "get", basically instructing your browser that the contents of this particular form should be submitted through an HTTP GET request. When set to "post", however, your browser will be instructed to send the information through an HTTP POST request. Instead of including all the form information as URL parameters, the POST HTTP request includes this input as part of the HTTP request body instead. Try this by pressing "Submit my information" while making sure that Chrome's Developer Tools are monitoring network requests. If you inspect the HTTP request, you'll notice that the request method is now set to "POST" and that Chrome includes an extra piece of information called "Form Data" to show you which information was included as part of the HTTP request body; see Figure 4-2. Note that the submitted information is now not embedded in the request URL.

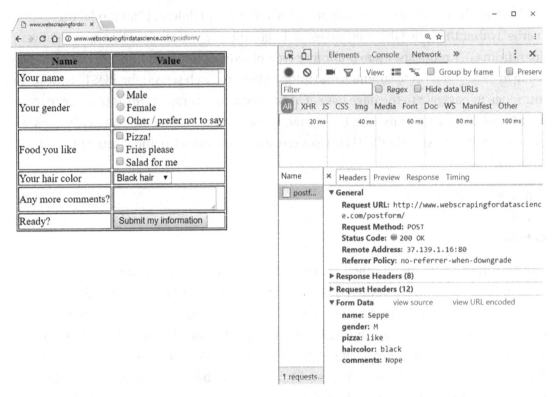

Figure 4-2. *Inspecting a POST HTTP request using Chrome Developer Tools*

Finally, try navigating to *http://www.webscrapingfordatascience.com/ postform2/*. This page works in exactly the same way as the one before, though with one notable difference: when you submit the form, the server doesn't send back the same contents as before, but provides an overview of the information that was just submitted. Since a web server is able to read out the HTTP request line containing the HTTP request method and URL, it can dynamically generate a different response depending on the type of response that was received, as well as undertake different actions. When a POST request comes in, for instance, the web server might decide to store the submitted information in a database before sending its reply.

A Word on Safety   Note that the URL displayed in the web browser hasn't changed, as the same URL can be used in combination with a POST or GET request. This shows how POST requests also offer a benefit in terms of safety: even when copying the URL above in an e-mail message, for instance, a new user opening this link will end up fetching the page using a GET request (and will hence not resubmit information to the site, but will just see a fresh form). It would be very unsafe if unsuspecting users could be tricked into clicking a URL that would immediately execute actions on their behalf. This doesn't mean that using POST requests fully protects you from nefarious actors, who can, for instance, instruct an unsuspecting visitor's web browser to perform a POST request using a piece of JavaScript code, but it does at least help a bit. This isn't to say that all web forms should use POST. Forms with search boxes, such as the one found on Google, for instance, are perfectly fine to use in combination with a GET request.

Finally, there is one more thing that we need to mention regarding web forms. In all the previous examples, the data submitted to the form was sent to the web server by constructing an HTTP request containing the same URL as the page the form was on, though this is not necessarily always the case. To indicate that a "submit" action should spawn a request to a different URL, the "action" attribute can be used as follows:

```
<form action="search.html">
  <!-- E.g. a search form -->
  <input type="text" name="query">
  <input type="submit" value="Search!">
</form>
```

This form snippet can be included on various pages with their according URLs, though submitting this form will lead to a page such as, for example, "search.html?query=Test".

Apart from GET and POST, there are some other HTTP methods to discuss, though these are far less commonly used, at least when surfing the web through a web browser. We'll take a look at these later; let us first summarize the two HTTP methods we know so far:

- GET: Used when typing in a URL in the address bar and pressing enter, clicking a link, or submitting a GET form. Here, the assumption is that the same request can be executed multiple times without "damaging" the user's experience. For instance, it's fine to refresh the URL "search.html?query=Test", but perhaps not the URL "doMoney Transfer?to=Bart&from=Seppe&amount=100".

- POST: Used when submitting a POST form. Here, the assumption is that this request will lead to an action being undertaken that should not be executed multiple times. Most browser will actually warn you if you try refreshing a page that resulted from a POST request ("you'll be resubmitting the same information again — are you sure this is what you want to do?").

Before moving on to other HTTP request methods, let's see how we can execute POST requests using Python. In case a web form is using a GET request to submit information, we've already seen how you can handle this use case simply by using the `requests.get` method with the `params` argument to embed the information as URL parameters. For a POST request, we just need to use a new method (`requests.post`) and a new `data` argument:

```python
import requests

url = 'http://www.webscrapingfordatascience.com/postform2/'

# First perform a GET request
r = requests.get(url)

# Followed by a POST request
formdata = {
    'name': 'Seppe',
    'gender': 'M',
    'pizza': 'like',
```

```
  'haircolor': 'brown',
  'comments': ''
  }
```

```
r = requests.post(url, data=formdata)
print(r.text)
```

Just like `params`, the `data` argument is supplied as a Python dictionary object representing name-value pairs. Take some time to play around with this URL in your web browser to see how data for the various input elements is actually submitted. In particular, note that for the radio buttons, they all bear the same "name" attribute to indicate that they belong to the same selection group, though the value ("M," "F," and "N" differs). If no selection is made at all, this element will not be included in the submitted form data. For the check boxes, note that they all have a different name, but that all their values are the same ("like"). If a check box is not selected, a web browser will not include the name-value pair in the submitted form data. Hence, some web servers don't even bother to check the value of such fields to determine whether they were selected, but just whether the name is present in the submitted data, though this can differ from web server to web server.

Duplicate Names   Some websites, such as those built using the PHP language, also allow us to define a series of check box elements with the same name attribute value. For sites built with PHP, you'll see these names ending with "[]", as in "food[]". This signposts to the web server that the incoming values should be treated as an array. On the HTTP side of things, this means that the same field name will end up multiple times in the request body. The same intricacy is present for URL parameters: technically, there is nothing preventing you from specifying multiple parameters with the same name and different values. The way how they'll be treated differs from server to server, though you might be wondering how we'd handle such a use case in requests, as both `params` and `data` are dictionary objects, which cannot contain the same keys twice. To work around this, both the `data` and `params` arguments allow to pass a list of "(name, value)" tuples to handle this issue.

For the "comments" field, note that it is included in the submitted form data even when nothing was filled in, in which case an empty value will be submitted. Again, in some cases you might as well leave out the name-value pair altogether, but it is up to the web server to determine how "picky" it wants to be.

Pickiness   Even submit buttons can be named, in which case they'll also end up in the submitted form data (with their value being equal to their "value" HTML attribute). Sometimes, web servers will use this information to determine which button was clicked in case more than one submit button is present in the same form and, in other cases, the name-value pair will end up being ignored. Just as was the case with URL parameters, web servers can differ widely in terms of how strict or flexible they are when interpreting requests. Nothing is stopping you to also put in other name-value pairs when submitting a POST request using requests, but again, the results might vary from site to site. Trying to match as closely what's happening in the browser is generally a good recommendation.

Note that in our example above, we're still "polite" in the sense that we first execute a normal GET request before sending the information through a POST, though this is not even required. We can simply comment out the requests.get line to submit our information. In some cases, however, web pages will be "smart" enough to prevent you from doing so. To illustrate this, try navigating to *http://www. webscrapingfordatascience.com/postform3/*, fill out, and submit the form. Now try the same again but wait a minute or two before pressing "Submit my information." The web page will inform you that "You waited too long to submit this information." Let's try submitting this form using requests:

```
import requests

url = 'http://www.webscrapingfordatascience.com/postform3/'

# No GET request needed?

formdata = {
    'name': 'Seppe',
    'gender': 'M',
    'pizza': 'like',
```

```
    'haircolor': 'brown',
    'comments': ''
    }

r = requests.post(url, data=formdata)

print(r.text)
# Will show: Are you trying to submit information from somewhere else?
```

This is strange: How does the web server know that we're trying to perform a POST request from Python in this case? The answer lies in one additional form element that is now present in the HTML source code (your value might differ):

```
<input type="hidden" name="protection" value="2c17abf5d5b4e326bea802600ff88405">
```

As can be seen, this form incorporates a new hidden field that will be submitted with the rest of the form data, conveniently named "protection." How about including it directly in our Python source code as follows:

```
import requests

url = 'http://www.webscrapingfordatascience.com/postform3/'

formdata = {
    'name': 'Seppe',
    'gender': 'M',
    'pizza': 'like',
    'haircolor': 'brown',
    'comments': '',
    'protection': '2c17abf5d5b4e326bea802600ff88405'
    }
r = requests.post(url, data=formdata)

print(r.text)
# Will show: You waited too long to submit this information. Try
<a href="./">again</a>.
```

Assuming you waited a minute before running this piece of code, the web server will now reply with a message indicating that it doesn't want to handle this request. Indeed, we can confirm (using our browser), that the "protection" field appears to change every time we refresh the page, seemingly randomly. To work our way around this, we have no other alternative but to first fetch out the form's HTML source using a GET request, get the value for the "protection" field, and then use that value in the subsequent POST request. By bringing in Beautiful Soup again, this can be easily done:

```python
import requests
from bs4 import BeautifulSoup

url = 'http://www.webscrapingfordatascience.com/postform3/'

# First perform a GET request
r = requests.get(url)

# Get out the value for protection
html_soup = BeautifulSoup(r.text, 'html.parser')
p_val = html_soup.find('input', attrs={'name': 'protection'}).get('value')

# Then use it in a POST request
formdata = {
    'name': 'Seppe',
    'gender': 'M',
    'pizza': 'like',
    'haircolor': 'brown',
    'comments': '',
    'protection': p_val
    }

r = requests.post(url, data=formdata)

print(r.text)
```

The example above illustrates a protective measure that you will encounter from time to time in real-life situations. Website administrators do not necessarily include such extra measures as a means to prevent web scraping (it makes a scraper's life a bit harder, but we've seen how we can work around it), but mainly for reasons of safety and improving the user experience. For instance, to prevent the same information being submitted twice (using the same "protection" value, for instance), or to prevent attacks where users are tricked to visit a certain web page on which a piece of JavaScript code will try to perform a POST request to another site, for instance, to initiate a money transfer or try to obtain sensitive information. Secure websites will hence often include such additional checks on their pages.

View State   There is one web server technology stack that is pretty famous for using such fields: Microsoft's ASP and ASP.NET. Websites built using this technology will — in the majority of cases — include a hidden input element in all forms with the name set to a cryptic "__VIEWSTATE", together with an encrypted value that can be very long. Not including this form element when trying to perform POST requests to sites built using this stack will lead to results not showing what you'd expect, and is oftentimes the first real-life annoyance web scrapers will run into. The solution is simple: just include these in your POST requests. Note that the resulting page sent in the HTTP reply might again contain such "__VIEWSTATE" element, so you'll have to make sure to fetch the value again and again to include it in every subsequent POST request.

There are a few more things worth mentioning before we can wrap up this section. First, you'll no doubt have noticed that we can now use a params and data argument, which look very similar. If GET requests use URL parameters, and POST requests send data as part of the HTTP request body, why do we need to separate arguments when we can already indicate the type of request, by using either the requests.get or request.post method? The answer lies in the fact that it is perfectly fine for an HTTP POST request to include both a request URL with parameters, as well as a request body containing form data. Hence, if you encounter the "<form>" tag definition in a page's source code:

```
<form action="submit.html?type=student" method="post">
[...]
</form>
```

You'll have to write the following in Python:

```
r = requests.post(url, params={'type': 'student'}, data=formdata)
```

You might also be wondering what would happen if we'd try to include the same information both in the URL parameters and in the form data:

```
import requests

url = 'http://www.webscrapingfordatascience.com/postform2/'

paramdata = {'name': 'Totally Not Seppe'}
formdata = {'name': 'Seppe'}
r = requests.post(url, params=paramdata, data=formdata)

print(r.text)
```

This particular web page will simply ignore the URL parameter and take into account the form data instead, but this is not necessarily always the case. Also, even although a "<form>" specifies "POST" as its "method" parameter, there might be rare cases where you can just as well submit this information as URL parameters instead using a simple GET request. These situations are rare, but they can happen. Nevertheless, the best recommendation is to stick as close as you can to the behavior that you observe by using the page normally.

Finally, there is one type of form element we haven't discussed before. Sometimes, you will encounter forms that allow you to upload files from your local machine to a web server:

```
<form action="upload.php" method="post" enctype="multipart/form-data">
  <input type="file" name="profile_picture">
  <input type="submit" value="Upload your profile picture">
</form>
```

Note the "file" input element in this source snippet, as well as the "enctype" parameter now present in the "<form>" tag. To understand what this parameter means, we need to talk a little bit about form encoding. Put simply, web forms will first "encode" the information contained in the form before embedding it in the HTTP POST request body. Currently, the HTML standard foresees three ways how this encoding can be done (which will end up as values for the "Content-Type" request header):

- application/x-www-form-urlencoded (the default): here, the request body is formatted similarly to what we've seen with URL parameters, hence the name "urlen-coded," that is, using ampersands ("&") and equals signs ("=") to separate data fields and name-value parts. Just an in URLs, certain characters should be encoded in a specific way, which requests will do automatically for us.

- text/plain: introduced by HTML 5 and generally only used for debugging purposes and hence extremely rare in real-life situations.

- multipart/form-data: this encoding method is significantly more complicated but allows us to include a file's contents in the request body, which might be in the form of binary, non-textual data — hence the need for a separate encoding mechanism.

As an example, consider an HTTP POST request with some request data included:

```
POST /postform2/ HTTP/1.1
Host: www.webscrapingfordatascience.com
Content-Type: application/x-www-form-urlencoded
[... Other headers]

name=Seppe&gender=M&pizza=like
```

Now consider an HTTP POST request with the request data encoded using "multipart/form-data":

```
POST /postform2/ HTTP/1.1
Host: www.webscrapingfordatascience.com
Content-Type: multipart/form-data; boundary=BOUNDARY
[... Other headers]

--BOUNDARY
Content-Disposition: form-data; name="name"
```

```
Seppe
--BOUNDARY
Content-Disposition: form-data; name="gender"

M
--BOUNDARY
Content-Disposition: form-data; name="pizza"

like
--BOUNDARY
Content-Disposition: form-data; name="profile_picture"; filename="me.jpg"
Content-Type: application/octet-stream

[... binary contents of me.jpg]
```

Definitely, the request body here looks more complicated, though we can see where the "multipart" moniker comes from: the request data is split up in multiple parts using a "boundary" string, which is determined (randomly, in most cases) by the request invoker. Luckily for us, we don't need to care too much about this when using requests. To upload a file, we simply use another argument, named `files` (which can be used together with the `data` argument):

```python
import requests

url = 'http://www.webscrapingfordatascience.com/postform2/'

formdata = {'name': 'Seppe'}
filedata = {'profile_picture': open('me.jpg', 'rb')}
r = requests.post(url, data=formdata, files=filedata)
```

The library will take care of setting the appropriate headers in the POST request (including picking a boundary) as well as encoding the request body correctly for you.

Multiple Files   For forms where multiple files can be uploaded, you'll mostly find that they utilize multiple "<input>" tags, each with a different name. Submitting multiple files then boils down to putting more key-value pairs in the `files` argument dictionary. The HTML standard also foresees a way to provide multiple files through one element only, using the "multiple" HTML parameter. To handle this in requests, you can pass a list to the files argument with each element being a tuple having two entries: the form field name, which can then appear multiple times throughout the list, and the file info — a tuple on its own containing the open call and other information about the file that's being sent. More information on this can be found under "POST Multiple Multipart-Encoded Files" in the requests documentation, though it is rather rare to encounter such upload forms in practice.

4.2 Other HTTP Request Methods

Now that we've seen how HTTP GET and POST requests work, we can take a brief moment to discuss the other HTTP methods that exist in the standard:

- GET: the GET method requests a representation of the specified URL. Requests using GET should only retrieve data and should have no other effect, such as saving or changing user information or perform other actions. In other words, GET requests should be "idempotent," meaning that it should be safe to execute the same request multiple times. Keep in mind that URL parameters can be included in the request URL (this is also the case for any other HTTP method), though GET requests can — technically — include an optional request body as well, but this is not recommended by the HTTP standard. As such, a web browser doesn't include anything in the request body when it performs GET requests, and they are not used by most (if not all) APIs either.

- POST: the POST method indicates that data is being submitted as part of a request to a particular URL, for example, a forum message, a file upload, a filled-in form, and so on. Contrary to GET, POST requests are not expected to be idempotent, meaning that submitting a POST request can bring about changes on the web server's end of things, such as, for example, updating your profile, confirming a money transaction, a purchase, and so on. A POST request encodes the submitted data as part of the request body.

- HEAD: the HEAD method requests a response just like the GET request does, but it indicates to the web server that it does not need to send the response body. This is useful in case you only want the response headers and not the actual response contents. HEAD requests cannot have a request body.

- PUT: the PUT method requests that the submitted data should be stored under the supplied request URL, thereby creating it if it does not exist already. Just as with a POST, PUT requests have a request body.

- DELETE: the DELETE method requests that the data listed under the request URL should be removed. The DELETE request does not have a request body.

- CONNECT, OPTIONS, TRACE, and PATCH: these are some less-commonly encountered request methods. CONNECT is generally used to request a web server to set up a direct TCP network connection between the client and the destination (web proxy servers will use this type of request), TRACE instructs the web server to just send the request back to the client (used for debugging to see if a middleman in the connection has changed your request somewhere in-between), OPTIONS requests the web server to list the HTTP methods it accepts for a particular URL (which might seem helpful, though is rarely used). PATCH finally allows us to request a partial modification of a specific resource.

Based on the above, it might seem that the set of HTTP methods corresponds pretty well to the basic set of SQL (Structured Query Language) commands, used to query and update information in relational databases, that is, GET to "SELECT" a resource given a URL, POST to "UPDATE" it, PUT to "UPSERT" it ("UPDATE" or "INSERT" if it does not exist), and DELETE to "DELETE" it. This being said, this is not how web browsers work. We've seen above that most web browsers will work using GET and POST requests only. That means that if you create a new profile on a social network site, for instance, your form data will simply be submitted through a POST request, and not a PUT. If you change your profile later on, another POST request is used. Even if you want to delete your profile, this action will be requested through a POST request.

This doesn't mean, however, that requests doesn't support these methods. Apart from `requests.get` and `requests.post`, you can also use the `requests.head`, `requests.put`, `requests.delete`, `requests.patch`, and `requests.options` methods.

A Word About APIs   Even although web browsers might only stick to GET and POST requests, there is a variety of networking protocols that put themselves on top of HTTP and do use the other request methods as well. Also, you'll find that many modern APIs — as offered by Facebook, Twitter, LinkedIn, and so on — also expose their functionality through HTTP and might use other HTTP request methods as well, a practice commonly referred to as REST (Representational State Transfer). It is then helpful to know that you can just use requests to access these as well. The difference between web scraping and using an API hence lies mainly in how structured the requests and replies are. With an API, you'll get back content in a structured format (such as XML or JSON), which can be easily parsed by computer programs. On the "regular" web, content is returned mainly as HTML formatted text. This is nice for human readers to work with after a web browser is done with it, but not very convenient for computer programs. Hence there is a need for something like Beautiful Soup. Note, however, that not all APIs are built on top of HTTP — some of them use other protocols, such as SOAP (Simple Object Access Protocol) as well, which then requires another set of libraries to access them.

4.3 More on Headers

Now that we're finished with an overview of HTTP request methods, it's time to take a closer look at another part of HTTP and how it comes into play when web scraping sites: the request headers. Up until now, we've been relying on requests to construct and send these headers for us. There are various cases, however, where we'll have to modify them ourselves.

Let's get started right away with the following example:

```python
import requests

url = 'http://www.webscrapingfordatascience.com/usercheck/'

r = requests.get(url)

print(r.text)
# Shows: It seems you are using a scraper

print(r.request.headers)
```

Note that the website responds with "It seems you are using a scraper." How does it know? When we open the same page in a normal browser, we see "Welcome, normal user," instead. The answer lies in the request headers that the requests library is sending:

```
{
  'User-Agent': 'python-requests/2.18.4',
  'Accept-Encoding': 'gzip, deflate',
  'Accept': '*/*',
  'Connection': 'keep-alive'
}
```

The requests library tries to be polite and includes a "User-Agent" header to announce itself. Of course, websites that want to prevent scrapers from accessing its contents can build in a simple check to block particular user agents from accessing their contents.

As such, we'll have to modify our request headers to "blend in," so to speak. In requests, sending custom headers is easily done through yet another argument: headers:

```
import requests

url = 'http://www.webscrapingfordatascience.com/usercheck/'

my_headers = {
  'User-Agent': 'Mozilla/5.0 (Windows NT 10.0; Win64; x64)
  AppleWebKit/537.36 ' + ' (KHTML, like Gecko) Chrome/61.0.3163.100
             Safari/537.36'
}

r = requests.get(url, headers=my_headers)

print(r.text)
print(r.request.headers)
```

This works. Note that the headers argument does not completely overwrite the default headers completely, but updates it instead, keeping around the default entries too.

Apart from the "User-Agent" header, there is another header that deserves special mention: the "Referer" header (originally a misspelling of referrer and kept that way since then). Browsers will include this header to indicate the URL of the web page that linked to the URL being requested. Some websites will check this to prevent "deep links" from working. To test this out, navigate to *http://www.webscrapingfordatascience. com/referercheck/* in your browser and click the "secret page" link. You'll be linked to another page (*http://www.webscrapingfordatascience.com/referercheck/secret. php*) containing the text "This is a totally secret page." Now try opening this URL directly in a new browser tab. You'll see a message "Sorry, you seem to come from another web page" instead. The same happens in requests:

```
import requests

url = 'http://www.webscrapingfordatascience.com/referercheck/secret.php'

r = requests.get(url)

print(r.text)
# Shows: Sorry, you seem to come from another web page
```

Try inspecting the requests your browser is making using your browser's developer tools and see if you can spot the "Referer" header being sent. You'll note that it says "http://www.webscrapingfordatascience.com/referercheck/" for the GET request to the secret page. When linked to from another website, or opened in a fresh tab, this referrer field will be different or not included in the request headers. Especially sites hosting image galleries will often resort to this tactic to prevent images being included directly in other web pages (they'd like images to only be visible from their own site and want to prevent paying for hosting costs for other pages using the images). When encountering such checks in requests, we can simply spoof the "Referer" header as well:

```
import requests

url = 'http://www.webscrapingfordatascience.com/referercheck/secret.php'

my_headers = {
    'Referer': 'http://www.webscrapingfordatascience.com/referercheck/'
}

r = requests.get(url, headers=my_headers)

print(r.text)
```

Just as we've seen at various occasions before, remember that web servers can get very picky in terms of headers that are being sent as well. Rare edge cases such as the order of headers, multiple header lines with the same header name, or custom headers being included in requests can all occur in real-life situations. If you see that requests is not returning the results you expect and have observed when using the site in your browser, inspect the headers through the developer tools to see exactly what is going on and duplicate it as well as possible in Python.

Duplicate Request and Response Headers   Just like the `data` and `params` arguments, `headers` can accept an `OrderedDict` object in case the ordering of the headers is important. Passing a list, however, is not permitted here, as the HTTP standard does not allow multiple request header lines bearing the same name. What is allowed is to provide multiple values for the same header by separating them with a comma, as in the line "Accept-Encoding: gzip, deflate". In that case, you can just pass the value as is with requests. However, that's not to say that some extremely weird websites or APIs might still use a setup where it deviates from the standard and checks for the same headers on multiple lines in the request. In that case, you'll have no choice but to implement a hack to extend requests. Note that response headers can contain multiple lines with the same name. Requests will automatically join them using a comma and put them under one entry when you access `r.headers`.

Finally, we should also take a closer look at the HTTP reply headers, starting first with the different HTTP response status codes. In most cases, the status code will be 200, the standard response for successful requests. The complete range of status codes can be categorized as follows:

- 1XX: informational status codes, indicating that a request was received and understood, but where the server indicates that the client should wait for an additional response. These are rarely encountered on the normal web.

- 2XX: success status codes, indicating that a request was received and understood, and processed successfully. The most prevalent status code here is 200 ("OK"), though 204 ("No Content" — indicating that the server will not return any content) and 206 ("Partial Content" — indicating that the server is delivering only a part of a resource, such as a video fragment) are sometimes used as well.

- 3XX: redirection status codes, indicating that the client must take additional action to complete the request, most often by performing a new request where the actual content can be found. 301 ("Moved Permanently"), for instance, indicates that this and all future requests should be directed to the given URL instead of the one used, 302

("Found") and 303 ("See Other") indicate that the response to the request can be found under another URL. 304 ("Not Modified") is used to indicate that the resource has not been modified since the version specified by the web browser in its cache-related headers, and that the browser can just reuse its previously downloaded copy. 307 ("Temporary Redirect") and 308 ("Permanent Redirect") indicate that a request should be repeated with another URL, either temporarily or permanently. More on redirects and working with them using requests later on.

- 4XX: client error status codes, indicating that an error occurred that was caused by the requester. The most commonly known status code here is 404 ("Not Found"), indicating that a requested resource could not be found but might become available later on. 410 ("Gone") indicates that a requested resource was available once but will not be available any longer. 400 ("Bad Request") indicates that the HTTP request was formatted incorrectly, 401 ("Unauthorized") is used to indicate that the requested resource is not available without authorization, whereas 403 ("Forbidden") indicates that the request is valid, including authentication, but that the user does not have the right credentials to access this resource. 405 ("Method Not Allowed") should be used to indicate that an incorrect HTTP request method was used. There's also 402 ("Payment Required"), 429 ("Too Many Requests"), and even 451 ("Unavailable For Legal Reasons") defined in the standard, though these are less commonly used.

- 5XX: server error status codes, indicating that the request appears valid, but that the server failed to process it. 500 ("Internal Server Error") is the most generic and most widely encountered status code in this set, indicating that there's perhaps a bug present in the server code or something else went wrong.

Who Needs Standards?   Even though there are a lot of status codes available to tackle a variety of different outcomes and situations, most web servers will not be too granular or specific in using them. It's hence not uncommon to get a 500 status code where a 400, 403, or 405 would have been more appropriate; or to get a 404 result code even when the page existed before and a 410 might be better. Also the different 3XX status codes are sometimes used interchangeably. As such, it's best not to overthink the definitions of the status codes and just see what a particular server is replying instead.

From the above listing, there are two topics that warrant a closer look: redirection and authentication. Let's take a closer look at redirection first. Open the page *http://www.webscrapingfordatascience.com/redirect/* in your browser. You'll see that you're immediately sent to another page ("destination.php"). Now do the same again while inspecting the network requests in your browser's developer tools (in Chrome, you should enable the "Preserve log" option to prevent Chrome from cleaning the log after the redirect happens). Note how two requests are being made by your browser: the first to the original URL, which now returns a 302 status code. This status code instructs your browser to perform a second request to the "destination.php" URL. How does the browser know what the URL should be? By inspecting the original URL's response, you'll note that there is now a "Location" response header present, which contains the URL to be redirected to. Note that we've also included another header in the HTTP response here: "SECRET-CODE," which we'll use in the Python examples later on. First, let's see how requests deals with redirection:

```
import requests

url = 'http://www.webscrapingfordatascience.com/redirect/'
r = requests.get(url)

print(r.text)
print(r.headers)
```

Note that we get the HTTP reply corresponding with the final destination ("you've been redirected here from another page!"). In most cases, this default behavior is quite helpful: requests is smart enough to "follow" redirects on its own when it receives 3XX status codes. But what if this is not what we want? What if we'd like to get the contents of the original page? This isn't shown in the browser either, but there might be a relevant response content present. What if we want to see the contents of the "Location" and "SECRET-CODE" headers manually? To do so, you can simply turn off requests default behavior of the following redirects through the `allow_redirects` argument:

```python
import requests

url = 'http://www.webscrapingfordatascience.com/redirect/'
r = requests.get(url, allow_redirects=False)

print(r.text)
print(r.headers)
```

Which will now show:

```
You will be redirected... bye bye!
{'Date': 'Fri, 13 Oct 2017 13:00:12 GMT',
 'Server': 'Apache/2.4.18 (Ubuntu)',
 'SECRET-CODE': '1234',
 'Location': 'http://www.webscrapingfordatascience.com/redirect/
              destination.php',
 'Content-Length': '34',
 'Keep-Alive': 'timeout=5, max=100',
 'Connection': 'Keep-Alive',
 'Content-Type': 'text/html; charset=UTF-8'}
```

There aren't many situations where you'll need to turn off redirect following, though it might be necessary in cases where you first wish to fetch the response headers (such as "SECRET-CODE") here before moving on. You'll then have to retrieve the "Location" header manually to perform the next `requests.get` call.

Redirects   Redirects using 3XX status codes are often used by websites, for instance, in the HTTP response following a POST request after data has been processed to send the browser to a confirmation page (for which it can then use a GET request). This is another measure taken to prevent users from submitting the same POST request twice in a row. Note that 3XX status codes are not the only way that browsers can be sent to another location. Redirect instructions can also be provided by means of a "<meta>" tag in an HTML document, which can include an optional timeout (web pages like these will often show something of the form "You'll be redirected after 5 seconds"), or through a piece of JavaScript code, which can fire off navigation instructions as well.

Finally, let's take a closer look at the 401 ("Unauthorized") status code, which seems to indicate that HTTP provides some sort of authentication mechanism. Indeed, the HTTP standard includes a number of authentication mechanisms, one of which can be seen by accessing the URL *http://www.webscrapingfordatascience.com/ authentication/*. You'll note that this site requests a username and password through your browser. If you press "Cancel," you'll note that the website responds with a 401 ("Unauthorized") result. Try refreshing the page and entering any username and password combination. The server will now respond with a normal 200 ("OK") reply. What actually goes on here is the following:

- Your browser performs a normal GET request to the page, and no authentication information is included.

- The website responds with a 401 reply and a "WWW-Authenticate" header.

- Your browser will take this as an opportunity to ask for a username and password. If "Cancel" is pressed, the 401 response is shown at this point.

- If the user provides a username and password, your browser will perform an additional GET request with an "Authorization" header included and the username and password encoded (though not really through a very strong encryption mechanism).

- The web server checks this request again, for example, to verify the
  username and password that were sent. If everything looks good, the
  server replies with a 200 page. Otherwise, a 403 ("Forbidden") is sent
  (if the password was incorrect, for instance, or the user doesn't have
  access to this page).

In requests, performing a request with a basic authentication is as simple as
including an "Authorization" header, so we still need to figure out how to encrypt the
username and password. Instead of doing this ourselves, requests provides another
means to do so, using the auth argument:

```
import requests

url = 'http://www.webscrapingfordatascience.com/authentication/'

r = requests.get(url, auth=('myusername', 'mypassword'))

print(r.text)
print(r.request.headers)
```

Apart from this basic authentication mechanism, which is pretty insecure (and
should only be used by websites in combination with HTTPS — otherwise your
information is transmitted using encryption that can be easily reversed), HTTP also
supports other "schemes" such as the digest-based authentication mechanism,
which requests supports as well. Although some older sites sometimes still use HTTP
authentication, you won't find this component of HTTP being used that often any longer.
Most sites will prefer to handle their authentication using cookies instead, which we'll
deal with in the next section.

4.4 Dealing with Cookies

All things considered, HTTP is a rather simple networking protocol. It is text based and
follows a simple request-and-reply-based communication scheme. In the simplest case,
every request-reply cycle in HTTP involves setting up a fresh new underlying network
connection as well, though the 1.1 version of the HTTP standard allows us to set up
"keep alive" connections, where a network connection is kept open for some period of
time so that multiple request-reply HTTP messages can be exchanged over the same
connection.

This simple request-reply-based approach poses some problems for websites, however. From a web server's point of view, every incoming request is completely independent of any previous ones and can be handled on its own. This is not, however, what users expect from most websites. Think for instance about an online shop where items can be added to a cart. When visiting the checkout page, we expect the web server to "remember" the items we selected and added previously. Similarly, when providing a username and password in a web form to access a protected page, the web server needs to have some mechanism to remember us, that is, to establish that an incoming HTTP request is related to a request that came in before.

In short, it didn't take long after the introduction of HTTP before a need arose to add a state mechanism on top of it, or, in other words, to add the ability for HTTP servers to "remember" information over the duration of a user's "session," in which multiple pages can be visited.

Note that based on what we've seen above, we can already think of some ways to add such functionality to a website:

- We could include a special identifier as a URL parameter that "links" multiple visits to the same user, for example, "checkout. html?visitor=20495".

- For POST requests, we could either use the same URL parameter, or include the "session" identifier in a hidden form field.

Some older websites indeed use such mechanisms, though this comes with several drawbacks:

- What happens if an unsuspecting user would copy the link and paste it in an e-mail? This would mean that another party opening this link will now be considered as the same user, and will be able to look through all their information.

- What happens if we close and reopen our browser? We'd have to log in again and go through all steps again as we're starting from a fresh session.

Linking Requests   Note that you might come up with other ways to link requests together as well. What about using the IP address (perhaps combined with the User-Agent) for a visiting user? Sadly, these approaches all come with similar security issues and drawbacks too. IP addresses can change, and it is possible for multiple computers to share the same public-facing IP address, meaning that all your office computers would appear as the same "user" to a web server.

To tackle this issue in a more robust way, two headers were standardized in HTTP in order to set and send "cookies," small textual bits of information. The way how this works is relatively straightforward. When sending an HTTP response, a web server can include "Set-Cookie" headers as follows:

```
HTTP/1.1 200 OK
Content-type: text/html
Set-Cookie: sessionToken=20495; Expires=Wed, 09 Jun 2021 10:10:10 GMT
Set-Cookie: siteTheme=dark
[...]
```

Note that the server is sending two headers here with the same name. Alternatively, the full header can be provided as a single line as well, where each cookie will be separated by a comma, as follows:

```
HTTP/1.1 200 OK
Content-type: text/html
Set-Cookie: sessionToken=20495; Expires=Wed, 09 Jun 2021 10:10:10 GMT,
siteTheme=dark
[...]
```

Capitalize on It   Some web servers will also use "set-cookie" in all lowercase to send back cookie headers.

The value of the "Set-Cookie" headers follows a well-defined standard:

- A cookie name and cookie value are provided, separated by an equals sign, "=". In the example above, for instance, the "sessionToken" cookie is set to "20495" and might be an identifier the server will use to recognize a subsequent page visit as belonging to the same session. Another cookie, called "siteTheme" is set to the value "dark," and might be used to store a user's preference regarding the site's color theme.

- Additional attributes can be specified, separated by a semicolon (";"). In the example above, an "Expires" attribute is set for the "sessionToken," indicating that a browser should store the cookie until the provided date. Alternatively, a "Max-Age" attribute can be used to gain a similar result. If none of these are specified, the browser will be instructed to remove the cookies once the browser window is closed.

Manual Deletion   Note that setting an "Expires" or "Max-Age" attribute should not be regarded as being a strict instruction. Users are free to delete cookies manually, for instance, or might simply switch to another browser or device as well.

- A "Domain" and "Path" attribute can be set as well to define the scope of the cookie. They essentially tell the browser what website the cookie belongs to and hence in which cases to include the cookie information in subsequent requests (more on this later on). Cookies can only be set on the current resource's top domain and its subdomains, and not for another domain and its subdomains, as otherwise websites would be able to control the cookies of other domains. If "Domain" and "Path" attributes are not specified by the server, they default to the domain and path of the resource that was requested.

- Finally, there are also the "Secure" and "HttpOnly" attributes, which do not come with an associated value. The "Secure" attribute indicates that the browser should limit communication of this cookie to encrypted transmissions (HTTPS). The "HttpOnly" attribute directs browsers not to expose cookies through channels other than HTTP (and HTTPS) requests. This means that the cookie cannot be accessed through, for example, JavaScript.

Secure Sessions   Note that care needs to be taken when defining a value for a session-related cookie such as "sessionToken" above. If this would be set to an easy-to-guess value, like a user ID or e-mail address, it would be very easy for malicious actors to simply spoof the value, as we'll see later on. Therefore, most session identifiers will end up being constructed randomly in a hard-to-guess manner. It is also good practice for websites to frequently expire session cookies or to replace them with a new session identifier from time to time, to prevent so-called "cookie hijacking": stealing another user's cookies to pretend that you're them.

When a browser receives a "Set-Cookie" header, it will store its information in its memory and will include the cookie information in all following HTTP requests to the website (provided the "Domain," "Path," "Secure," and "HttpOnly" checks pass). To do so, another header is used, this time in the HTTP request, simply named "Cookie":

```
GET /anotherpage.html HTTP/1.1
Host: www.example.com
Cookie: sessionToken=20495; siteTheme=dark
[...]
```

Note that here, the cookie names and values are simple included in one header line, and are separated by a semicolon (";"), not a comma as is the case for other multi-valued headers. The web server is then able to parse these cookies on its end, and can then derive that this request belongs to the same session as a previous one, or do other things with the provided information (such as determine which color theme to use).

Evil Cookies   Cookies are an essential component for the modern web to work, but they have gotten a bad reputation over the past years, especially after the EU Cookie Directive was passed and cookies were mentioned in news articles as a way for social networks to track you across the Internet. In themselves, cookies are in fact harmless, as they can be sent only to the server setting them or a server in the same domain. However, a web page may contain images or other components stored on servers in other domains, and to fetch those, browsers will send the cookies belonging to those domains as well in the request. That is, you might be visiting a page on "`www.example.com`," to which only cookies belonging to that domain will be sent, but that site might host an image coming from another website, such as "`www.facebook.com/image.jpg`." To fetch this image, a new request will be fired off, now including Facebook's cookies. Such cookies are called "third-party cookies," and are frequently used by advertisers and others to track users across the Internet: if Facebook (or advertisers) instruct the original site to set the image URL to something like "`www.facebook.com/image.jpg?i_came_from=www-example-org`," it will be able to stitch together the provided information and determine which users are visiting which sites. Many privacy activists have warned against the use of such cookies, any many browser vendors have built-in ways to block sending such cookies.

Fingerprinting   Because of the increasing backlash against third-party cookies, many publishers on the web have been looking for other means to track users. JSON Web Tokens, IP addresses, ETag headers, web storage, Flash, and many other approaches have been developed to either set information in a browser that can be retrieved later on, so that users can be remembered; or to "fingerprint," a device and browser in such a way that the fingerprint is unique across the whole visitor population and can also be used as a unique identifier. Some particularly annoying approaches will use a combination of various techniques to set "evercookies," which are particularly hard to wipe from a device. Not surprisingly, browser vendors continue to implement measures to prevent such practices.

Let's now go over some examples to learn how we can deal with cookies in requests. The first example we'll explore can be found at *http://www.webscrapingfordatascience.com/cookielogin/*. You'll see a simple login page. After successfully logging in (you can use any username and password in this example), you'll be able to access a secret page over at the website *http://www.webscrapingfordatascience.com/cookielogin/secret.php*. Try closing and reopening your browser (or just open an Incognito or Private Mode browser tab) and accessing the secret URL directly. You'll see that the server detects that you're not sending the right cookie information and blocks you from seeing the secret code. The same can be observed when trying to access this page directly using requests:

```
import requests

url = 'http://www.webscrapingfordatascience.com/cookielogin/secret.php'

r = requests.get(url)

print(r.text)
# Shows: Hmm... it seems you are not logged in
```

Obviously, we need to set and include a cookie. To do so, we'll use a new argument, called cookies. Note that we could use the headers argument (which we've seen before) to include a "Cookie" header, but we'll see that cookies is a bit easier to use, as requests will take care of formatting the header appropriately. The question is now where to get the cookie information from. We could fall back on our browser's developer tools, and get the cookie from the request headers there and include it as follows:

```
import requests

url = 'http://www.webscrapingfordatascience.com/cookielogin/secret.php'

my_cookies = {'PHPSESSID': 'ijfatbjege43lnsfn2b5c37706'}

r = requests.get(url, cookies=my_cookies)

print(r.text)
# Shows: This is a secret code: 1234
```

However, if we'd want to use this scraper later on, this particular session identifier might have been flushed and become invalid.

PHPSESSID   We use the PHP scripting language to power our examples, so that the cookie name to identify a user's session is named "PHPSESSID". Other websites might use "session," "SESSION_ID," "session_id," or any other name as well. Do note, however, that the value representing a session should be constructed randomly in a hard-to-guess manner. Simply setting a cookie "is_logged_in=true" or "logged_in_user=Seppe" would of course be very easy to guess and spoof.

We hence need to resort to a more robust system as follows: we'll first perform a POST request simulating a login, get out the cookie value from the HTTP response, and use it for the rest of our "session." In requests, we can do this as follows:

```python
import requests

url = 'http://www.webscrapingfordatascience.com/cookielogin/'

# First perform a POST request
r = requests.post(url, data={'username': 'dummy', 'password': '1234'})

# Get the cookie value, either from
# r.headers or r.cookies print(r.cookies)
my_cookies = r.cookies

# r.cookies is a RequestsCookieJar object which can also
# be accessed like a dictionary. The following also works:
my_cookies['PHPSESSID'] = r.cookies.get('PHPSESSID')

# Now perform a GET request to the secret page using the cookies
r = requests.get(url + 'secret.php', cookies=my_cookies)

print(r.text)
# Shows: This is a secret code: 1234
```

This works, though there are some real-life cases where you'll have to deal with more complex login (and cookie) flows. Navigate to the next example over at *http://www. webscrapingfordatascience.com/redirlogin/*. You'll see the same login page again, but note that you're now immediately redirected to the secret page after successfully logging in. If you use the same Python code as in the fragment above, you'll note that you're not able to log in correctly and that the cookies being returned from the POST request are empty. The reason behind this is related to something we've seen before: requests will automatically follow HTTP redirect status codes, but the "Set-Cookie" response header is present in the response following the HTTP POST request, and not in the response for the redirected page. We'll hence need to use the `allow_redirects` argument once again:

```python
import requests

url = 'http://www.webscrapingfordatascience.com/redirlogin/'

# First perform a POST request -- do not follow the redirect
r = requests.post(url, data={'username': 'dummy', 'password': '1234'},
                  allow_redirects=False)

# Get the cookie value, either from r.headers or r.cookies
print(r.cookies)

my_cookies = r.cookies

# Now perform a GET request manually to the secret page using the cookies
r = requests.get(url + 'secret.php', cookies=my_cookies)

print(r.text)
# Shows: This is a secret code: 1234
```

As a final example, navigate to *http://www.webscrapingfordatascience.com/ trickylogin/*. This site works in more or less the same way (explore it in your browser), though note that the "<form>" tag now includes an "action" attribute. We might hence change our code as follows:

```python
import requests

url = 'http://www.webscrapingfordatascience.com/trickylogin/'

# First perform a POST request -- do not follow the redirect
# Note that the ?p=login parameter needs to be set
r = requests.post(url, params={'p': 'login'},
                  data={'username': 'dummy', 'password': '1234'},
                  allow_redirects=False)
```

```
# Set the cookies
my_cookies = r.cookies

# Now perform a GET request manually to the secret page using the cookies
r = requests.get(url, params={'p': 'protected'}, cookies=my_cookies)

print(r.text)
# Hmm... where is our secret code?
```

This doesn't seem to work for this example. The reason for this is that this particular example also checks whether we've actually visited the login page, and are hence not only trying to directly submit the login information. In other words, we need to add in another GET request first:

```
import requests

url = 'http://www.webscrapingfordatascience.com/trickylogin/'

# First perform a normal GET request to get the form
r = requests.post(url)

# Then perform the POST request -- do not follow the redirect
r = requests.post(url, params={'p': 'login'},
                  data={'username': 'dummy', 'password': '1234'},
                  allow_redirects=False)

# Set the cookies
my_cookies = r.cookies

# Now perform a GET request manually to the secret page using the cookies
r = requests.get(url, params={'p': 'protected'}, cookies=my_cookies)

print(r.text)
# Hmm... still no secret code?
```

This also does not seem to work yet. Let's think about this for a second... Obviously, the way that the server would "remember" that we've seen the login screen is by setting a cookie, so we need to retrieve that cookie after the first GET request to get the session identifier at that moment:

```
import requests

url = 'http://www.webscrapingfordatascience.com/trickylogin/'

# First perform a normal GET request to get the form
r = requests.post(url)

# Set the cookies already at this point!
my_cookies = r.cookies

# Then perform the POST request -- do not follow the redirect
# We already need to use our fetched cookies for this request!
r = requests.post(url, params={'p': 'login'},
                  data={'username': 'dummy', 'password': '1234'},
                  allow_redirects=False,
                  cookies=my_cookies)

# Now perform a GET request manually to the secret page using the cookies
r = requests.get(url, params={'p': 'protected'}, cookies=my_cookies)

print(r.text)
# Still no secret?
```

Again, this fails... the reason for this (you can verify this as well in your browser) is that this site changes the session identifier after logging in as an extra security measure.

The following code shows what happens — and finally gets out our secret code:

```
import requests

url = 'http://www.webscrapingfordatascience.com/trickylogin/'

# First perform a normal GET request to get the form
r = requests.post(url)

# Set the cookies
my_cookies = r.cookies
print(my_cookies)
```

```
# Then perform the POST request -- do not follow the redirect
# Use the cookies we got before
r = requests.post(url, params={'p': 'login'},
                  data={'username': 'dummy', 'password': '1234'},
                  allow_redirects=False,
                  cookies=my_cookies)

# We need to update our cookies again
# Note that the PHPSESSID value will have changed
my_cookies = r.cookies
print(my_cookies)

# Now perform a GET request manually to the secret page
# using the updated cookies
r = requests.get(url, params={'p': 'protected'}, cookies=my_cookies)

print(r.text)
# Shows: Here is your secret code: 3838.
```

The above examples show a simple truth about dealing with cookies, which should not sound surprising now that we now how they work: every time an HTTP response comes in, we should update our client-side cookie information accordingly. In addition, we need to be careful when dealing with redirects, as the "Set-Cookie" header might be "hidden" inside the original HTTP response, and not in the redirected page's response. This is quite troublesome and will indeed quickly lead to messy scraping code, though fear not, as requests provides another abstraction that makes all of this much more straightforward: sessions.

4.5 Using Sessions with Requests

Let's immediately jump in an introduce requests' sessions mechanism. Our "tricky login" example above can simply be rewritten as follows:

```
import requests

url = 'http://www.webscrapingfordatascience.com/trickylogin/'

my_session = requests.Session()
```

```
r = my_session.post(url)
r = my_session.post(url, params={'p': 'login'},
                    data={'username': 'dummy', 'password': '1234'})
r = my_session.get(url, params={'p': 'protected'})

print(r.text)
# Shows: Here is your secret code: 3838.
```

You'll notice a few things going on here: first, we're creating a `requests.Session` object and using it to perform HTTP requests, using the same methods (get, post) as above. The example now works, without us having to worry about redirects or dealing with cookies manually.

This is exactly what the requests' session mechanism aims to offer: basically, it specifies that various requests belong together — to the same session — and that requests should hence deal with cookies automatically behind the scenes. This is a huge benefit in terms of user friendliness, and makes requests shine compared to other HTTP libraries in Python. Note that sessions also offers an additional benefit apart from dealing with cookies: if you need to set global header fields, such as the "User-Agent" header, this can simply be done once instead of using the `headers` argument every time to make a request:

```
import requests

url = 'http://www.webscrapingfordatascience.com/trickylogin/'

my_session = requests.Session()
my_session.headers.update({'User-Agent': 'Chrome!'})

# All requests in this session will now use this User-Agent header:

r = my_session.post(url)
print(r.request.headers)

r = my_session.post(url, params={'p': 'login'},
                    data={'username': 'dummy', 'password': '1234'})
print(r.request.headers)

r = my_session.get(url, params={'p': 'protected'})
print(r.request.headers)
```

Even if you think a website is not going to perform header checks or use cookies, it is still a good idea to create a session nonetheless and use that.

Clearing Cookies   If you ever need to "clean" a session by clearing its cookies, you can either set up a new session, or simply call:

```
my_session.cookies.clear()
```

This works since `RequestsCookieJar` objects (which represent a collection of cookies in requests) behave like normal Python dictionaries.

4.6 Binary, JSON, and Other Forms of Content

We're almost done covering everything requests has to offer. There are a few more intricacies we need to discuss, however. So far, we've only used requests to fetch simple textual or HTML-based content, though remember that to render a web page, your web browser will typically fire off a lot of HTTP requests, including requests to fetch images. Additionally, files, like a PDF file, say, are also downloaded using HTTP requests.

PDF Scraping   In what follows, we'll show you how to download files, though it might be interesting to know that "PDF scraping" is an interesting area on its own. You might be able to set up a scraping solution using requests to download a collection of PDF files, though extracting information from such files might still be challenging. However, several tools have been developed to also help you in this task, which are out of scope here. Take a look at the "PDFMiner" and "slate" libraries, for instance, to extract text, or "tabula-py," to extract tables. If you're willing to switch to Java, "PDF Clown" is an excellent library to work with PDF files as well. Finally, for those annoying PDF files containing scanned images, OCR software such as "Tesseract" might come in handy to automate your data extraction pipeline as well.

To explore how this works in requests, we'll be using an image containing a lovely picture of a kitten at *http://www.webscrapingfordatascience.com/files/kitten.jpg*. You might be inclined to just use the following approach:

```
import requests

url = 'http://www.webscrapingfordatascience.com/files/kitten.jpg'
r = requests.get(url)

print(r.text)
```

However, this will not work and leave you with a "UnicodeEncodeError." This is not too unexpected: we're downloading binary data now, which cannot be represented as Unicode text. Instead of using the text attribute, we need to use content, which returns the contents of the HTTP response body as a Python bytes object, which you can then save to a file:

```
import requests

url = 'http://www.webscrapingfordatascience.com/files/kitten.jpg'
r = requests.get(url)

with open('image.jpg', 'wb') as my_file:
    my_file.write(r.content)
```

Don't Print   It's not a good idea to print out the r.content attribute, as the large amount of text may easily crash your Python console window.

However, note that when using this method, Python will store the full file contents in memory before writing it to your file. When dealing with huge files, this can easily overwhelm your computer's memory capacity. To tackle this, requests also allows to stream in a response by setting the stream argument to True:

```
import requests

url = 'http://www.webscrapingfordatascience.com/files/kitten.jpg'

r = requests.get(url, stream=True)
# You can now use r.raw
# r.iter_lines
# and r.iter_content
```

Once you've indicated that you want to stream back a response, you can work with the following attributes and methods:

- `r.raw` provides a file-like object representation of the response. This is not often used directly and is included for advanced purposes.

- The `iter_lines` method allows you to iterate over a content body line by line. This is handy for large textual responses.

- The `iter_content` method does the same for binary data.

Let's use `iter_content` to complete our example above:

```python
import requests

url = 'http://www.webscrapingfordatascience.com/files/kitten.jpg'

r = requests.get(url, stream=True)

with open('image.jpg', 'wb') as my_file:
    # Read by 4KB chunks
    for byte_chunk in r.iter_content(chunk_size=4096):
        my_file.write(byte_chunk)
```

There's another form of content you'll encounter a lot when working with websites: JSON (JavaScript Object Notation), a lightweight textual data interchange format that is both relatively easy for humans to read and write and easy for machines to parse and generate. It is based on a subset of the JavaScript programming language, but its usage has become so widespread that virtually every programming language is able to read and generate it. You'll see this format used a lot by various web APIs these days to provide content messages in a structured way. There are other data interchange formats as well, such as XML and YAML, though JSON is by far the most popular one.

As such, it is interesting to know how to deal with JSON-based requests and response messages, not only when planning to use requests to access a web API that uses JSON, but also in various web scraping situations. To explore an example, head over to *http://www.webscrapingfordatascience.com/jsonajax/*. This page shows a simple lotto number generator. Open your browser's developer tools, and try pressing the

"Get lotto numbers" button a few times... By exploring the source code of the page, you'll notice a few things going on:

- Even though there's a button on this page, it is not wrapped by a "<form>" tag.

- When pressing the button, part of the page is updated without completely reloading the page.

- The "Network" tab in Chrome will show that HTTP POST requests are being made when pressing the button.

- You'll notice a piece of code in the page source wrapped inside "<script>" tags.

This page uses JavaScript (inside the "<script>" tags) to perform so-called AJAX requests. AJAX stands for Asynchronous JavaScript And XML and refers to the use of JavaScript to communicate with a web server. Although the name refers to XML, the technique can be used to send and receive information in various formats, including JSON, XML, HTML, and simple text files. AJAX's most appealing characteristic lies in its "asynchronous" nature, meaning that it can be used to communicate with a web server without having to refresh a page completely. Many modern websites use AJAX to, for example, fetch new e-mails, fetch notifications, update a live news feed, or send data, all without having to perform a full page refresh or submit a form. For this example, we don't need to focus too much on the JavaScript side of things, but can simply look at the HTTP requests it is making to see how it works:

- POST requests are being made to "results.php".

- The "Content-Type" header is set to "application/x-www-form-urlencoded," just like before. The client-side JavaScript will make sure to reformat a JSON string to a an encoded equivalent.

- An "api_code" is submitted in the POST request body.

- The HTTP response has a "Content-Type" header set to "application/json," instructing the client to interpret the result as JSON data.

Working with JSON-formatted replies in requests is easy as well. We can just use text as before and, for example, convert the returned result to a Python structure manually (Python provides a json module to do so), but requests also provides a helpful json method to do this in one go:

```
import requests

url = 'http://www.webscrapingfordatascience.com/jsonajax/results.php'

r = requests.post(url, data={'api_code': 'C123456'})

print(r.json())
print(r.json().get('results'))
```

There's one important remark here, however. Some APIs and sites will also use an "application/json" "Content-Type" for formatting the request and hence submit the POST data as plain JSON. Using a requests' data argument will not work in this case. Instead, we need to use the json argument, which will basically instruct requests to format the POST data as JSON:

```
import requests

url = 'http://www.webscrapingfordatascience.com/jsonajax/results2.php'

# Use the json argument to encode the data as JSON:
r = requests.post(url, json={'api_code': 'C123456'})

# Note the Content-Type header in the request:
print(r.request.headers)

print(r.json())
```

Internal APIs   Even if the website you wish to scrape does not provide an API, it's always recommended to keep an eye on your browser's developer tools networking information to see if you can spot JavaScript-driven requests to URL endpoints that return nicely structured JSON data. Even although an API might not be documented, fetching the information directly from such "internal APIs" is always a good idea, as this will avoid having to deal with the HTML Soup.

This concludes our in-depth discussion of HTTP. In the next chapter, we continue with JavaScript. With what we've seen now, you have a strong tool set available to deal with sites, even if they use JavaScript to perform asynchronous HTTP requests, as we have just seen. However, JavaScript can do a lot more and for sites where JavaScript is being used to change a website's content, set and check cookies, or check whether a real browser is accessing a site, you'll still encounter cases where replicating all this behavior through requests and Beautiful Soup becomes too cumbersome. In these cases, we'll have no choice but to simulate a full browser in order to scrape websites.

CHAPTER 5

Dealing with JavaScript

Together with HTML and CSS, JavaScript forms the third and final core building block of the modern web. We've already seen JavaScript appearing occasionally throughout this book, and it's time that we take a closer look at it. As we'll soon see in this chapter, our sturdy requests plus Beautiful Soup combination is no longer a viable approach to scrape JavaScript-heavy pages. As such, this chapter will introduce you to another library, Selenium, which is used to automate and control a full-fledged web browser.

5.1 What Is JavaScript?

Put briefly, JavaScript — just like Python — is a high-level programming language. Contrary to many other programming languages, its core functionality lies in making web pages more interactive and dynamic. Many websites use it to do so, and all modern web browsers support it right out of the box through a built-in JavaScript engine. In the earlier days, JavaScript was implemented solely in client-side web browsers and only geared to be used in that context. However, it's interesting to note that JavaScript as a language has garnered a lot of interest over recent years, and can now be found running server-side programs as well, even including complete web servers, or as the engine behind various desktop applications.

Java and JavaScript   It's kind of becoming a moot point these days, but it's still worth mentioning in passing that although there are some similarities between JavaScript and Java, the two programming languages are really distinct and differ quite a lot.

© Seppe vanden Broucke and Bart Baesens 2018
S. vanden Broucke and B. Baesens, *Practical Web Scraping for Data Science*,
https://doi.org/10.1007/978-1-4842-3582-9_5

JavaScript code can be found inside of "<script>" tags in an HTML document, either verbatim:

```
<script type="text/javascript">
// JavaScript code comes here
</script>
```

Or by setting a "src" attribute referring to the location where the code can be found:

```
<script type="text/javascript" src="my_file.js"></script>
```

We're not going to learn how to write JavaScript here, but we will need to figure out how we can deal with the language when trying to scrape websites that intensively use it. Let's get started by trying to work with JavaScript using the tools we've seen so far.

5.2 Scraping JavaScript

Navigate to the URL at *http://www.webscrapingfordatascience.com/ simplejavascript/*. This simple web page shows three random quotes, but it uses JavaScript to do so. Inspect the source code of the page, and you'll find the following JavaScript fragment there:

```
<script>
  $(function() {
  document.cookie = "jsenabled=1";
    $.getJSON("quotes.php", function(data) {
      var items = [];
      $.each(data, function(key, val) {
        items.push("<li id='" + key + "'>" + val + "</li>");
      });
      $("<ul/>", {
        html: items.join("")
        }).appendTo("body");
      });
    });
</script>
```

Inspecting Elements versus View Source, Again   This is a good moment to reemphasize the difference between viewing a page's source and inspecting elements using your browser's developer tools. The "View source" option shows the HTML code as it was returned by the web server, and it will contain the same contents as r.text when using requests. Inspecting elements, on the other hand, provides a "cleaned up" version after the HTML was parsed by your web browser and provides a live and dynamic view. That's why you can inspect the quote elements, but do not see them appear in the page's source, as they're loaded by JavaScript after the page's contents have been retrieved by the browser.

This JavaScript fragment does the following:

- To code is wrapped in a "$()" function. This is not part of standard JavaScript, but instead a mechanism provided by jQuery, a popular JavaScript library that is loaded using another "<script>" tag. The code defined in the function will be executed once the browser is finished with loading the page.

- The code inside the function starts by setting a "jsenabled" cookie. Indeed, JavaScript is able to set and retrieve cookies as well.

- Next, a "getJSON" function is used to perform another HTTP request to fetch the quotes, which are added by inserting a "<ul>" tag in the "<body>".

Let's now take a look at how we can deal with this use case using requests and Beautiful Soup:

```
import requests
from bs4 import BeautifulSoup

url = 'http://www.webscrapingfordatascience.com/simplejavascript/'

r = requests.get(url)

html_soup = BeautifulSoup(r.text, 'html.parser')

# No tag will be found here
ul_tag = html_soup.find('ul')
print(ul_tag)
```

```
# Show the JavaScript code
script_tag = html_soup.find('script', attrs={'src': None})
```

print(script_tag)

As you will observe, the contents of the page are just returned as is, but neither requests nor Beautiful Soup come with a JavaScript engine included, meaning that no JavaScript will be executed, and no "<ul>" tag will be found on the page. We can take a look at the "<script>" tag, but to Beautiful Soup, this will look like any other HTML tag with a bunch of text inside. We have no way to parse and query the actual JavaScript code.

In simple situations such as this one, this is not necessarily a problem. We know that the browser is making requests to a page at "quotes.php", and that we need to set a cookie. We can still scrape the data directly:

import requests

url = 'http://www.webscrapingfordatascience.com/simplejavascript/quotes.php'

```
# Note that cookie values need to be provided as strings
r = requests.get(url, cookies={'jsenabled': '1'})
```

print(r.json())

This works (try to see what happens without setting the cookie), though on more complex pages, coming up with such an approach might be rather daunting. Some websites will go quite far to prevent you from "reverse engineering" what is happening in the JavaScript code. As an example, head over to *http://www.webscrapingfordatascience.com/complexjavascript/*. You'll note that this page loads additional quotes by scrolling to the bottom of the list. Inspecting the "<script>" tags now shows the following obfuscated mess:

```
<script>
var _0x110b=["","\x6A\x6F\x69\x6E","\x25","\x73[...]\x6C"];
function sc(){
var _0xe9a7=["\x63\x6F\x6F\x6B\x69\x65","\x6E\x6F\x6E\x63\x65\x3D2545"];
document[_0xe9a7[0]]= _0xe9a7[1]
}$(function(){sc();
```

```
function _0x593ex2(_0x593ex3){
return decodeURIComponent([...],
function(_0x593ex4){
return _0x110b[2]+ [...]}
$(_0x110b[16])[_0x110b[15]]({
padding:20,
nextSelector:_0x110b[9],
contentSelector:_0x110b[0],
callback:function(){
$(_0x110b[14])[_0x110b[13]](function(_0x593ex5){
$(this)[_0x110b[10]](_0x593ex2($(this)[_0x110b[10]]()));
$(this)[_0x110b[12]](_0x110b[11])})}})})
</script>
```

It's obviously unfeasible to figure out what is going on here. For your web browser, interpreting and running this code might be simple, but to us humans, it is not. Luckily, we can still try to inspect the network requests to figure out what is happening here, to some extent:

- Requests are made once again to a "quotes.php" page with a "p" URL parameter, used for pagination.

- Two cookies are used here: "nonce" and "PHPSESSID." The latter we've encountered before, and is simply included in the "Set-Cookie" response header for the main page. The "nonce" cookie, however, is not, which indicates that it might be set through JavaScript, though we don't really know where (clever readers might be able to figure it out, but stay with us for the sake of the example).

Untangling JavaScript   Security researchers will oftentimes try to untangle obfuscated JavaScript code such as the one seen here to figure out what a particular piece of code is doing. Obviously, this is a daunting and exhausting task.

We can at least try to "steal" the nonce's cookie value to see if we can get anything using requests (note that your "nonce" cookie value might be different):

```
import requests

url = 'http://www.webscrapingfordatascience.com/complexjavascript/'

my_session = requests.Session()

# Get the main page first to obtain the PHPSESSID cookie
r = my_session.get(url)

# Manually set the nonce cookie
my_session.cookies.update({
    'nonce': '2315'
    })

r = my_session.get(url + 'quotes.php', params={'p': '0'})

print(r.text)
# Shows: No quotes for you!
```

Sadly, this doesn't work. Figuring out why requires some creative thinking, though we can take a guess at what might be going wrong here. We're getting a fresh session identifier by visiting the main page as if we were coming from a new browsing session to provide the "PHPSESSID" cookie. However, we're reusing the "nonce" cookie value that our browser was using. The web page might see that this "nonce" value does not match with the "PHPSESSID" information. As such, we have no choice but to also reuse the "PHPSESSID" value. Again, yours might be different (inspect your browser's network requests to see which values it is sending for your session):

```
import requests

url = 'http://www.webscrapingfordatascience.com/complexjavascript/'

my_cookies = {
    'nonce': '2315',
    'PHPSESSID': 'rtc4l3m3bgmjo2fqmi0og4nv24'
    }

r = requests.get(url + 'quotes.php', params={'p': '0'}, cookies=my_cookies)

print(r.text)
```

This leaves us with a different result, at least:

```
<div class="quote decode">TGlmZ[...]EtydXNlCg==</div>
<div class="quote decode">CVdoY[...]iBIaWxsCg==</div>
<div class="quote decode">CVNOc[...]Wluc3RlaW4K</div>
<br><br><br><br>
<a class="jscroll-next" href="quotes.php?p=3">Load more quotes</a>
```

This looks like HTML containing our quotes, but note that every quote seems to be encoded in some way. Obviously, there must be another piece of programming logic on JavaScript's side to decode the quotes and display them after fetching them, though once again, it is very hard to figure out how exactly this is done by inspecting the JavaScript source.

In addition, we're reusing the "nonce" and "PHPSESSID" cookies by getting them from a browser. Once the web server decides that it hasn't heard from us for some time and we retry to run our script with the same values, the web server will refuse to answer once again as we're using outdated cookies. We'd then need to reload the page in the browser and replace the cookie values in our script above.

Cracking the Code   Admittedly, you might have already recognized the encoding scheme that is being used here to encode the quotes: base64, and might hence be able to decode them within Python as well (doing so is easy). This is fine, of course, and in a real-life setting this might be an approach to consider, though note that the example above just wants to show that there might be a point at which trying to reverse engineer JavaScript becomes relatively hard. Similarly, you might have figured out where the "nonce" cookie value is hidden in the JavaScript code above, and would be able to fetch it using some regular expressions with Python, though again this is besides the argument we're trying to make here.

Obviously, this approach comes with a number of issues, which — sadly — we're unable to solve using what we've seen so far. The solution to this problem is easy to describe: we're seeing the quotes appear in our browser window, which is executing JavaScript, so can't we get them out from there? Indeed, for sites making heavy use of JavaScript, we'll have no choice but to emulate a full browser stack, and to move away from requests and Beautiful Soup. This is exactly what we'll do in the next section, using another Python library: Selenium.

Are You a Browser?   Some web companies will also utilize JavaScript in order to check visitors to see whether they're using a real browser before showing the contents of a page. CloudFlare, for instance, calls this its "Browser Integrity Check." In these cases, it will also be extremely hard to reverse engineer its workings to pretend you're a browser when using requests.

5.3 Scraping with Selenium

Selenium is a powerful web scraping tool that was originally developed for the purpose of automated website testing. Selenium works by automating browsers to load a website, retrieve its contents, and perform actions like a user would when using the browser. As such, it's also a powerful tool for web scraping. Selenium can be controlled from various programming languages, such as Java, C#, PHP, and of course, Python.

It's important to note that Selenium itself does not come with its own web browser. Instead, it requires a piece of integration software to interact with a third party, called a WebDriver. WebDrivers exist for most modern browsers, including Chrome, Firefox, Safari, and Internet Explorer. When using these, you'll see a browser window open up on your screen and perform the actions you have specified in your code.

Headless Mode   In what follows, we'll install and use the Chrome WebDriver together with Selenium, but feel free to use another WebDriver if you prefer. It's also interesting to know that although it can be helpful to watch what the browser is doing when creating a web scraper, this comes with an additional overhead and might be harder to set up on servers with no display attached. Not to worry, as Selenium also provides WebDrivers for so-called "headless" browsers, which run "invisibly" without showing a graphical user interface. If you want to learn more, take a look at the PhantomJS WebDriver. PhantomJS is a "headless" browser written in JavaScript and is often used on its own for website testing and scraping (in JavaScript), but can — through its WebDriver — also be used as an engine in Selenium and hence, Python. Note that just as any other modern web browser, PhantomJS will be able to render HTML, work with cookies, and execute JavaScript behind the scenes.

Just as was the case with requests and Beautiful Soup, installing Selenium itself is simple with pip (refer back to section 1.2.1 if you still need to set up Python 3 and pip):

```
pip install -U selenium
```

Next, we need to download a WebDriver. Head to *https://sites.google.com/a/ chromium.org/chromedriver/downloads* and download the latest release file matching your platform (Windows, Mac, or Linux). The ZIP file you downloaded will contain an executable called "chromedriver.exe" on Windows or just "chromedriver" otherwise. The easiest way to make sure Selenium can see this executable is simply by making sure it is located in the same directory as your Python scripts, in which case the following small example should work right away:

```
from selenium import webdriver

url = 'http://www.webscrapingfordatascience.com/complexjavascript/'

driver = webdriver.Chrome()
driver.get(url)

input('Press ENTER to close the automated browser')
driver.quit()
```

If you prefer to keep the WebDriver executable somewhere else, it is also possible to pass its location as you construct the Selenium webdriver object in Python like so (however, we'll assume that you keep the executable in the same directory for the examples that follow to keep the code a bit shorter):

```
driver_exe = 'C:/Users/Seppe/Desktop/chromedriver.exe'
# If you copy-paste the path with back-slashes, make sure to escape them
# E.g.: driver_exe = 'C:\\Users\\Seppe\\Desktop\\chromedriver.exe'
driver = webdriver.Chrome(driver_exe)
```

If you run this first example, you'll notice that a new Chrome window appears, with the warning that this window is being controlled by automated test software; see Figure 5-1. The Python script waits until you press enter, and then quits the browser window.

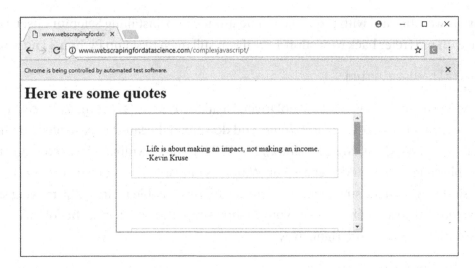

Figure 5-1. *A Chrome window being controlled by Selenium*

Once a page is loaded, you'll want to be able to get out HTML elements, just as we did with Beautiful Soup before. Selenium's list of methods looks slightly different, but shouldn't be too difficult based on what you've learned so far:

- `find_element_by_id`

- `find_element_by_name`

- `find_element_by_xpath`

- `find_element_by_link_text`

- `find_element_by_partial_link_text`

- `find_element_by_tag_name`

- `find_element_by_class_name`

- `find_element_by_css_selector`

Note that all of these also come with an `find_elements_by_*` variant (note the "s"), which — just like the `find_all` method in Beautiful Soup — return a list of elements (represented as `WebElement` objects in Selenium) instead of the first matching element. Also important to note is that the methods above will raise a `NoSuchElementException` exception in case an element could not be found. Beautiful Soup simply returns None in this case. Most of the methods in this case will be self-explanatory, though some require further explanation.

The find_element_by_link_text method selects elements by matching its inner text. The find_element_by_partial_link_text method does the same, but partially matches the inner text. These come in handy in a lot of cases, for instance, to find a link based on its textual value.

The find_element_by_name method selects elements based on the HTML "name" attribute, whereas find_element_by_tag_name uses the actual tag name.

The find_element_by_css_selector method is similar to Beautiful Soup's select method, but comes with a more robust CSS selector rule parser.

The find_element_by_xpath method does something similar. XPath is a language used for locating nodes in XML documents. As HTML can be regarded as an implementation of XML (also referred to as XHTML in this case), Selenium can use the XPath language to select elements. XPath extends beyond the simple methods of locating by id or name attributes. A quick overview of common XPath expressions is given below:

- nodename selects all nodes with the name "nodename";

- / selects from the root node;

- // can be used to skip multiple levels of nodes and search through all descendants to perform a selection;

- . selects the current node;

- .. selects the parent of the current node;

- @ selects attributes.

These expressions can be chained together to form powerful rules. Some example XPath rules are:

- /html/body/form[1]: get the first form element inside the "<body>" tag inside the "<html>" tag.

- //form[1]: get the first form element inside the document.

- //form[@id='my_form']: get the form element with an "id" attribute set to "my_form".

Inspect Mode   If you want to use CSS selectors or XPath expressions in your scrapers, remember that you can right-click HTML element in the Elements tab of Chrome's Developer Tools and select "Copy, Copy selector" and "Copy XPath" to give you an outline of what an appropriate selector or expression might look like.

Let's modify our script to showcase these methods, for instance, to get out the quotes' contents:

```python
from selenium import webdriver

url = 'http://www.webscrapingfordatascience.com/complexjavascript/'

# chromedriver should be in the same path as your Python script
driver = webdriver.Chrome()
driver.get(url)

for quote in driver.find_elements_by_class_name('quote'):
    print(quote.text)

input('Press ENTER to close the automated browser')
driver.quit()
```

Let's take a step back here and see what's going on: after navigating to our page, we use the find_elements_by_class_name method to retrieve quote elements. For each such quote element (a WebElement object), we use its text attribute to print out its text. Just as with Beautiful Soup's find and find_all, it's possible to call the find_element_* and find_elements_* methods on retrieved elements in case you want to "dig deeper."

Sadly, running this code doesn't seem to work, as no quotes are displayed at all. The reason for this is because our browser will take some time — even if only half a second — to execute the JavaScript, fetch the quotes, and display them. Meanwhile, our Python script is already hard at work to try to find quote elements, which at that moment are not yet there. We might simply slap in a sleep line in our code to wait a few seconds, but Selenium comes with a more robust approach: wait conditions.

Selenium provides two types of waits: implicit and explicit. An implicit wait makes WebDriver poll the page for a certain amount of time every time when trying to locate an element. Think of the implicit wait as a "catch all" where we wait every time when trying to locate an element up to a specified amount of time. By default, the implicit waiting time is set to zero, meaning that Selenium will not wait at all. It's easy to change this, however. As such, the following code works:

```
from selenium import webdriver

url = 'http://www.webscrapingfordatascience.com/complexjavascript/'

driver = webdriver.Chrome()

# Set an implicit wait
driver.implicitly_wait(10)

driver.get(url)

for quote in driver.find_elements_by_class_name('quote'):
    print(quote.text)

input('Press ENTER to close the automated browser')
driver.quit()
```

Implicit waits are helpful when you're just getting started with Selenium, but explicit waits offer more fine-grained control. An explicit wait makes the WebDriver wait for a certain, given condition to return a non-False value before proceeding further with execution. This condition will be tried over and over again until it returns something, or until a given timeout has elapsed. To use explicit waits, we rely on the following imports:

```
from selenium.webdriver.common.by import By
from selenium.webdriver.support.ui import WebDriverWait
from selenium.webdriver.support import expected_conditions as EC
```

The EC object above comes with a large number of built-in conditions, so we don't need to code these ourselves. Out of the box, Selenium provides the following conditions:

- alert_is_present: checks whether an alert is present.
- element_located_selection_state_to_be(locator, is_selected): checks whether an element is located matching a locator (see explanation below) and its selection state matches is_selected (True of False).

- `element_located_to_be_selected(locator)`: checks whether an element (a `WebElement` object) is located matching a `locator` (see explanation below) and is selected.

- `element_selection_state_to_be(element, is_selected)`: checks whether the selection state of an `element` (a WebElement object) matches `is_selected` (True or False).

- `element_to_be_selected(element)`: checks whether an `element` (a `WebElement` object) is selected.

- `element_to_be_clickable(locator)`: checks whether an element is located matching a `locator` (see explanation below) and can be clicked (i.e., is enabled).

- `frame_to_be_available_and_switch_to_it(locator)`: checks whether a frame matching a `locator` (see explanation below) is located and can be switched to, once found, the condition switches to this frame.

- `invisibility_of_element_located(locator)`: checks whether an element matching a `locator` (see explanation below) is invisible or not present on the page (visibility means that the element is not only displayed or has a height and width that is greater than 0).

- `new_window_is_opened(current_handles)`: checks whether a new window has opened.

- `number_of_windows_to_be(num_windows)`: checks whether a specific number of windows have opened.

- `presence_of_all_elements_located(locator)`: checks whether there is at least one element present on the page matching a `locator` (see explanation below). If found, the condition returns a list of matching elements.

- `presence_of_element_located(locator)`: checks whether there is at least one element present on the page matching a `locator` (see explanation below). If found, the condition returns the first matching element.

- `staleness_of(element)`: checks whether an element (a `WebElement` object) has been removed from the page.

- `text_to_be_present_in_element(locator, text_)`: checks whether a given string is present in an element matching a `locator` (see explanation below).

- `text_to_be_present_in_element_value(locator, text_)`: checks whether a given string is present in an element's value attribute matching a `locator` (see explanation below).

- `title_contains(title)`: checks whether the title of the page contains the given string.

- `title_is(title)`: checks whether the title of the page is equal to the given string.

- `url_changes(url)`: checks whether the URL is different from a given one.

- `url_contains(url)`: checks whether the URL contains the given one.

- `url_matches(pattern)`: checks whether the URL matches a given regular expression pattern.

- `url_to_be(url)`: checks whether the URL matches the given one.

- `visibility_of(element)`: checks whether a present element (a `WebElement` object) is visible (visibility means that the element is not only displayed but also has a height and width that is greater than 0).

- `visibility_of_all_elements_located(locator)`: checks whether all elements matching a `locator` (see explanation below) are also visible. If this is the case, returns a list of matching elements.

- `visibility_of_any_elements_located(locator)`: checks whether any element matching a `locator` (see explanation below) is visible. If this is the case, returns the first visible element.

- `visibility_of_element_located(locator)`: checks whether the first element matching a `locator` (see explanation below) is also visible. If this is the case, return the element.

Let's see how they work by modifying our example to use an explicit wait:

```python
from selenium import webdriver
from selenium.webdriver.common.by import By
from selenium.webdriver.support.ui import WebDriverWait
from selenium.webdriver.support import expected_conditions as EC

url = 'http://www.webscrapingfordatascience.com/complexjavascript/'

driver = webdriver.Chrome()

driver.get(url)

quote_elements = WebDriverWait(driver, 10).until(
    EC.presence_of_all_elements_located(
        (By.CSS_SELECTOR, ".quote:not(.decode)")
    )
)

for quote in quote_elements:
    print(quote.text)

input('Press ENTER to close the automated browser')
driver.quit()
```

This example works as follows. First, we create a WebDriverWait object using our WebDriver and a given amount of seconds we'd like to wait for it. We then call the until method on this object, to which we need to provide a condition object, the predefined presence_of_all_elements_located in our case. Note that this condition requires a locator argument, which is simply a tuple containing two elements. The first element determines based on which method an element should be selected (possible options are: By.ID, By.XPATH, By.NAME, By.TAG_NAME, By.CLASS_NAME, By.CSS_SELECTOR, By.LINK_TEXT, and By.PARTIAL_LINK_TEXT); and the second element provides the actual value. Here, our locator states that we want to find elements by a given CSS selector rule, specifying all elements with a "quote" CSS class but not with a "decode" CSS class, as we want to wait until the JavaScript code on the page is done decoding the quotes.

This condition will be checked over and over again until 10 seconds have passed, or until the condition returns something that is not False, that is, the list of matching elements in the case of presence_of_all_elements_located. We can then directly loop over this list and retrieve the quotes' contents.

Note that locator tuples can be used outside of conditions as well. That is, instead of using the fine-grained `find_element_by_*` and `find_elements_by_*` methods as discussed above, you can also use the general-purpose `find_element` and `find_elements` methods and supply a By argument and the actual value, for example, as in "driver. find_elements(By.XPATH, '//a')". This is a bit less readable in plain code, but comes in handy in you want to define your own custom conditions, which is possible as well as the following example shows:

```python
class at_least_n_elements_found(object):
    def __init__(self, locator, n):
        self.locator = locator
        self.n = n

    def __call__(self, driver):
        # Do something here and return False or something else
        # Depending on whether the condition holds
        elements = driver.find_elements(*self.locator)
        if len(elements) >= self.n:
            return elements
        else:
            return False

wait = WebDriverWait(driver, 10)
element = wait.until(
        at_least_n_elements_found((By.CLASS_NAME, 'my_class'), 3)
)
```

In the example, a new condition is defined that is similar to `presence_of_all_elements_located`, but waits until at least a given amount of matching elements could be found.

There's still one more thing we need to solve. So far, our example only returns the first three quotes. We still need to figure out a way to scroll down in our list of quotes using Selenium in order to load all of them. To do so, Selenium comes with a selection of "actions" that can be performed by the browser, such as clicking elements, clicking and dragging, double-clicking, right-clicking, and so on, which we could use in order to move down the scroll bar. However, since we're interacting with a full-fledged browser, we can also instruct it to execute JavaScript commands. Since it is possible to scroll using

JavaScript as well, we can use the execute_script method in order to send a JavaScript command to the browser.

The following code fragments shows this in action, using our custom wait condition:

```python
from selenium import webdriver
from selenium.webdriver.common.by import By
from selenium.webdriver.support.ui import WebDriverWait
from selenium.webdriver.support import expected_conditions as EC
from selenium.common.exceptions import TimeoutException

class at_least_n_elements_found(object):
    def __init__(self, locator, n):
        self.locator = locator
        self.n = n

    def __call__(self, driver):
        elements = driver.find_elements(*self.locator)
        if len(elements) >= self.n:
            return elements
        else:
            return False

url = 'http://www.webscrapingfordatascience.com/complexjavascript/'

driver = webdriver.Chrome()
driver.get(url)

# Use an implicit wait for cases where we don't use an explicit one
driver.implicitly_wait(10)

div_element = driver.find_element_by_class_name('infinite-scroll')
quotes_locator = (By.CSS_SELECTOR, ".quote:not(.decode)")

nr_quotes = 0
while True:
    # Scroll down to the bottom
    driver.execute_script(
        'arguments[0].scrollTop = arguments[0].scrollHeight',
        div_element)
```

```
    # Try to fetch at least nr_quotes+1 quotes
    try:
        all_quotes = WebDriverWait(driver, 3).until(
            at_least_n_elements_found(quotes_locator, nr_quotes + 1)
        )
    except TimeoutException as ex:
        # No new quotes found within 3 seconds, assume this is all there is
        print("... done!")
        break
    # Otherwise, update the quote counter
    nr_quotes = len(all_quotes)
    print("... now seeing", nr_quotes, "quotes")

# all_quotes will contain all the quote elements
print(len(all_quotes), 'quotes found\n')
for quote in all_quotes:
    print(quote.text)

input('Press ENTER to close the automated browser')
driver.quit()
```

Execute this code and follow what's happening in the browser along with the output
you obtain in the Python console, which should show the following:

```
... now seeing 3 quotes
... now seeing 6 quotes
... now seeing 9 quotes
... now seeing 12 quotes
... now seeing 15 quotes
... now seeing 18 quotes
... now seeing 21 quotes
... now seeing 24 quotes
... now seeing 27 quotes
... now seeing 30 quotes
... now seeing 33 quotes
... done!
33 quotes found
```

Life is about making an impact, not making an income. -Kevin Kruse
Whatever the mind of man can conceive and believe, it can achieve. ↵
-Napoleon Hill
Strive not to be a success, but rather to be of value. -Albert Einstein
Two roads diverged in a wood, and —II took the one less traveled by, And ↵
that has made all the difference. -Robert Frost
[...]

If you'd like to see how this would work without using JavaScript commands and actions instead, you can take a look at the following fragment (note the two new imports). In the next section, we'll talk more about interacting with a web page through actions.

```python
from selenium import webdriver
from selenium.webdriver.common.by import By
from selenium.webdriver.support.ui import WebDriverWait
from selenium.webdriver.support import expected_conditions as EC
from selenium.common.exceptions import TimeoutException
from selenium.webdriver.common.action_chains import ActionChains
from selenium.webdriver.common.keys import Keys

class at_least_n_elements_found(object):
    def __init__(self, locator, n):
        self.locator = locator
        self.n = n

    def __call__(self, driver):
        elements = driver.find_elements(*self.locator)
        if len(elements) >= self.n:
            return elements
        else:
            return False

url = 'http://www.webscrapingfordatascience.com/complexjavascript/'

driver = webdriver.Chrome()
driver.get(url)

# Use an implicit wait for cases where we don't use an explicit one
driver.implicitly_wait(10)
```

```python
div_element = driver.find_element_by_class_name('infinite-scroll')
quotes_locator = (By.CSS_SELECTOR, ".quote:not(.decode)")

nr_quotes = 0
while True:
    # Scroll down to the bottom, now using action (chains)
    action_chain = ActionChains(driver)
    # Move to our quotes block
    action_chain.move_to_element(div_element)
    # Click it to give it focus
    action_chain.click()
    # Press the page down key about 10 ten times
    action_chain.send_keys([Keys.PAGE_DOWN for i in range(10)])
    # Do these actions
    action_chain.perform()

    # Try to fetch at least nr_quotes+1 quotes
    try:
        all_quotes = WebDriverWait(driver, 3).until(
            at_least_n_elements_found(quotes_locator, nr_quotes + 1)
        )
    except TimeoutException as ex:
        # No new quotes found within 3 seconds, assume this is all there is
        print("... done!")
        break
    # Otherwise, update the quote counter
    nr_quotes = len(all_quotes)
    print("... now seeing", nr_quotes, "quotes")

# all_quotes will contain all the quote elements
print(len(all_quotes), 'quotes found\n')
for quote in all_quotes:
    print(quote.text)

input('Press ENTER to close the automated browser')
driver.quit()
```

5.4 More on Selenium

Now that we have seen how to find elements, work with conditions and waits in Selenium, and send JavaScript commands to the browser, it's time that we dig a little bit deeper in the library. To do so, we'll switch back to one of our earlier examples at *http://www.webscrapingfordatascience.com/postform2/* — the form over there will provide us with a good playground to explore Selenium's capabilities.

Let's start by talking a bit more about navigation. We have already seen the get method to navigate to a URL using Selenium. Similarly, you can also call a driver's forward and back methods (these take no arguments) to go forward and backward in the browser's history. Regarding cookies, it is helpful to know that — since Selenium uses a real browser — we don't need to worry about cookie management ourselves. If you want to output the cookies currently available, you can call the get_cookies method on a WebDriver object. The add_cookie method allows you to set a new cookie (it expects a dictionary with "name" and "value" keys as its argument).

Let's start with a fresh new Selenium program to work with our web form. We'll just use implicit waiting here to keep things straightforward:

```
from selenium import webdriver

url = 'http://www.webscrapingfordatascience.com/postform2/'

driver = webdriver.Chrome()
driver.implicitly_wait(10)

driver.get(url)

input('Press ENTER to close the automated browser')
driver.quit()
```

Every time you retrieve elements using the find_element_by_* and find_elements_by_* methods (or the general-purpose find_element and find_elements methods), Selenium will return WebElement objects. There are a number of interesting methods and attributes you can access for such objects:

- First, keep in mind that you can use the find_element_by_* and find_elements_by_* methods (or the general-purpose find_element and find_elements methods) again on a WebElement object to start a new search for an element starting from the current element, just like we could in Beautiful Soup. This is helpful to retrieve nested elements in the page.

- The `click` method clicks the element.

- The `clear` method clears the text of the element if it's a text entry element.

- The `get_attribute` method gets an HTML attribute or property of the element, or None. For example, `element.get_attribute('href')`.

- The `get_property` method gets a property of the element.

- The `is_displayed`, `is_enabled` and `is_selected` methods return a Boolean value indicating whether an element is visible to the user, enabled, or selected respectively. The latter is used to see whether a check box or radio button is selected.

- The `send_keys` method simulates typing into the element. It takes either a string, a list, or a series of keystrokes passed as a list.

- The `submit` method submits a form.

- The `value_of_css_property` method returns the value of a CSS property of the element for a given property name.

- The `screenshot` method saves a screenshot of the current element to a given PNG file name.

- The `location` attribute provides the location of the element, `size` returns the size of the element, and `rect` returns a dictionary with both the size and location. The `parent` attribute references the `WebDriver` instance this element was found from; `tag_name` contains the element's tag name; and the `text` attribute returns the text of the element, which we've already used before. The `page_source` attribute of a `WebElement` object will give back the live HTML source of the full page.

Getting HTML   The page_source attribute of a WebElement object will return the live HTML source of the full page. If you only want to get the HTML source for the element only, however, you need to access the "innerHTML" attribute using element.get_attribute('innerHTML') The "outerHTML" attribute does the same but includes the element's tags themselves as well. These attributes can be useful if you still want to use, for example, Beautiful Soup to parse some components of the page.

To work with drop-downs ("<select>" tags), Selenium allows you to construct a Select object (found under "selenium.webdriver.support.select") based on a given WebElement object. Once you have created such an object, you can call the following methods or access the following attributes:

- select_by_index(index): select the option at the given index.

- select_by_value(value): select all options that have a value matching the argument.

- select_by_visible_text(text): select all options that display text matching the argument.

- The methods above all come with deselect_* variants as well to deselect options. The deselect_all method clears all selected entries (note that the select tag can support multiple selections).

- all_selected_options: returns a list of all selected options belonging to this select tag.

- first_selected_option: the first selected option in this select tag (or the currently selected option in a normal select that only allows for a single selection).

- options: returns a list of all options belonging to this select tag.

Let's use what we've seen so far to start filling in our form:

```
from selenium import webdriver
from selenium.webdriver.support.select import Select
from selenium.webdriver.common.keys import Keys
```

```
url = 'http://www.webscrapingfordatascience.com/postform2/'

driver = webdriver.Chrome()
driver.implicitly_wait(10)

driver.get(url)

driver.find_element_by_name('name').send_keys('Seppe')
driver.find_element_by_css_selector('input[name="gender"][value="M"]').
click()
driver.find_element_by_name('pizza').click()
driver.find_element_by_name('salad').click()
Select(driver.find_element_by_name('haircolor')).select_by_value('brown')
driver.find_element_by_name('comments').send_keys(
    ['First line', Keys.ENTER, 'Second line'])

input('Press ENTER to submit the form')

driver.find_element_by_tag_name('form').submit()
# Or: driver.find_element_by_css_selector('input[type="submit"]').click()

input('Press ENTER to close the automated browser')
driver.quit()
```

Take some time to run this example and go through the code. Note the use of the special Keys helper object that can be used to send keys such as ENTER, PAGE DOWN, and so on.

Instead of working with actions directly as seen above, Selenium also provides an ActionChains object (found under "selenium.webdriver.common.action_chains") to construct more fine-grained chains of actions. This is useful for doing more complex actions like hover over and drag and drop. They're constructed by providing a driver object, and expose the following methods:

- click(on_element=None): clicks an element. If None is given, uses the current mouse position.

- click_and_hold(on_element=None): holds down the left mouse button on an element or the current mouse position.

- release(on_element=None): releasing a held mouse button on an element or the current mouse position.

- `context_click(on_element=None)`: performs a context click (right-click) on an element or the current mouse position.

- `double_click(on_element=None)`: double-clicks an element or the current mouse position.

- `move_by_offset(xoffset, yoffset)`: move the mouse to an offset from current mouse position.

- `move_to_element(to_element)`: move the mouse to the middle of an element.

- `move_to_element_with_offset(to_element, xoffset, yoffset)`: move the mouse by an offset of the specified element. Offsets are relative to the top-left corner of the element.

- `drag_and_drop(source, target)`: holds down the left mouse button on the source element, then moves to the target element and releases the mouse button.

- `drag_and_drop_by_offset(source, xoffset, yoffset)`: holds down the left mouse button on the source element, then moves to the target offset and releases the mouse button.

- `key_down(value, element=None)`: sends a keypress only, without releasing it. Should only be used with modifier keys (i.e., Control, Alt, and Shift).

- `key_up(value, element=None)`: releases a modifier key.

- `send_keys(*keys_to_send)`: sends keys to current focused element.

- `send_keys_to_element(element, *keys_to_send)`: sends keys to an element.

- `pause(seconds)`: wait for a given amount of seconds.

- `perform()`: performs all stored actions defined on the action chain. This is normally the last command you'll give to a chain.

- `reset_actions()`: clears actions that are already stored on the remote end.

The following example is functionally equivalent to the above, but it uses action chains to fill in most of the form fields:

```
from selenium import webdriver
from selenium.webdriver.support.select import Select
from selenium.webdriver.common.keys import Keys
from selenium.webdriver.common.action_chains import ActionChains

url = 'http://www.webscrapingfordatascience.com/postform2/'

driver = webdriver.Chrome()
driver.implicitly_wait(10)

driver.get(url)

chain = ActionChains(driver)
chain.send_keys_to_element(driver.find_element_by_name('name'), 'Seppe')
chain.click(driver.find_element_by_css_selector('input[name="gender"]
[value="M"]'))
chain.click(driver.find_element_by_name('pizza'))
chain.click(driver.find_element_by_name('salad'))
chain.click(driver.find_element_by_name('comments'))
chain.send_keys(['This is a first line', Keys.ENTER, 'And this a second'])
chain.perform()

Select(driver.find_element_by_name('haircolor')).select_by_value('brown')

input('Press ENTER to submit the form')

driver.find_element_by_tag_name('form').submit()
# Or: driver.find_element_by_css_selector('input[type="submit"]').click()

input('Press ENTER to close the automated browser')
driver.quit()
```

We've now discussed the most important parts of the Selenium library. Take some time to read through the documentation of the library over at *http://selenium-python. read thedocs.io/*. Note that working with Selenium can take some getting used to. Since we're using a full browser stack, you'll no doubt have noted that Selenium's actions correspond with UI actions as if they would be performed by a human user (e.g., clicking, dragging, selecting), instead of directly working with HTTP and HTML as we have done before. The final chapter includes some larger, real-life projects with Selenium as well. It is good to keep in mind that, for the majority of projects, requests and Beautiful Soup will be just fine to write web scrapers. It is only when dealing with complex, or highly interactive pages, that you'll have to switch to Selenium. That said, keep in mind that you can still use Beautiful Soup with Selenium as well in case you need to parse a particular element retrieved with Selenium to get out the information.

From Web Scraping to Web Crawling

So far, the examples in the book have been quite simple in the sense that we only scraped (mostly) a single page. When writing web scrapers, however, there are many occasions where you'll wish to scrape multiple pages and even multiple websites. In this context, the name "web crawler" is oftentimes used, as it will "crawl" across a site or even the complete web. This chapter illustrates how you can go from writing web scrapers to more elaborate web crawlers, and it highlights important aspects to keep in mind when writing a crawler. You'll also learn how you can store results in a database for later access and analysis.

6.1 What Is Web Crawling?

The difference between "web scraping" and "web crawling" is relatively vague, as many authors and programmers will use both terms interchangeably. In general terms, the term "crawler" indicates a program's ability to navigate web pages on its own, perhaps even without a well-defined end goal or purpose, endlessly exploring what a site or the web has to offer. Web crawlers are heavily used by search engines like Google to retrieve contents for a URL, examine that page for other links, retrieve the URLs for those links, and so on…

When writing web crawlers, there are some subtle design choices that come into play that will change the scope and nature of your project:

- In many cases, crawling will be restricted to a well-defined set of pages, for example, product pages of an online shop. These cases are relatively easy to handle, as you're staying within the same domain and have an expectation about what each product page will look like, or about the types of data you want to extract.

© Seppe vanden Broucke and Bart Baesens 2018
S. vanden Broucke and B. Baesens, *Practical Web Scraping for Data Science*,
https://doi.org/10.1007/978-1-4842-3582-9_6

- In other cases, you will restrict yourself to a single website (a single domain name), but do not have a clear target regarding information extraction in mind. Instead, you simply want to create a copy of the site. In such cases, manually writing a scraper is not an advisable approach. There are many tools available (for Windows, Mac, and Linux) that will help you to make an offline copy of a website, including lots of configurable options. To find these, look for "website mirror tool."

- Finally, in even more extreme cases, you'll want to keep your crawling very open ended. For example, you might wish to start from a series of keywords, Google each of them, crawl to the top 10 results for every query, and crawl those pages for, say, images, tables, articles, and so on. Obviously, this is the most advanced use case to handle.

Writing a robust crawler requires that you put various checks in place and think carefully about the design of your code. Since crawlers can end up on any part of the web and will oftentimes run for long amounts of time, you'll need to think carefully about stopping conditions, keeping track of which pages you visited before (and whether it's already time to visit them again), how you will store the results, and how you can make sure a crashed script can be restarted without losing its current progress. The following overview provides some general best practices and food for thought:

- **Think carefully about which data you actually want to gather:** Can you extract what you need by scraping a set of predefined websites, or do you really need to discover websites you don't know about yet? The first option will always lead to easier code in terms of writing and maintenance.

- **Use a database:** It's best to use a database to keep track of links to visit, visited links, and gathered data. Make sure to timestamp everything so you know when something was created and last updated.

- **Separate crawling from scraping:** Most robust crawlers separate the "crawling" part (visiting websites, extracting links, and putting them in a queue, that is, gathering the pages you wish to scrape) from the actual "scraping" part (extracting information from pages). Doing both in one and the same program or loop is quite error prone.

In some cases, it might be a good idea to have the crawler store a complete copy of a page's HTML contents so that you don't need to revisit it once you want to scrape out information.

- **Stop early:** When crawling pages, it's always a good idea to incorporate stopping criteria right away. That is, if you can already determine that a link is not interesting at the moment of seeing it, don't put it in the "to crawl" queue. The same applies when you scrape information from a page. If you can quickly determine that the contents are not interesting, then don't bother continuing with that page.

- **Retry or abort:** Note that the web is a dynamic place, and links can fail to work or pages can be unavailable. Think carefully about how many times you'll want to retry a particular link.

- **Crawling the queue:** That said, the way you deal with your queue of links is important as well. If you just apply a simple FIFO (first in first out) or LIFO (last in first out) approach, you might end up retrying a failing link in quick succession, which might not be what you want to do. Building in cooldown periods is hence important as well.

- **Parallel programming:** In order to make your program efficient, you'll want to write it in such a way that you can spin up multiple instances that all work in parallel. Hence the need for a database-backed data store as well. Always assume that your program might crash at any moment and that a fresh instance should be able to pick up the tasks right away.

- **Keep in mind the legal aspects of scraping:** In the next chapter, we take a closer look at legal concerns. Some sites will not be too happy about scrapers visiting them. Also make sure that you do not "hammer" a site with a storm of HTTP requests. Although many sites are relatively robust, some might even go down and will be unable to serve normal visitors if you're firing hundreds of HTTP requests to the same site.

We're not going to create a Google competitor in what follows, but we will give you a few pointers regarding crawling in the next section, using two self-contained examples.

6.2 Web Crawling in Python

As our first example, we're going to work with the page at *http://www.webscrapingfordatascience.com/crawler/*. This page tries to emulate a simple numbers station, and it will basically spit out a random list of numbers and links to direct you to a different page.

Numbers Stations   This example was inspired by a hilarious experiment from 2005 that can be seen at *http://www.drunkmenworkhere.org/218*. The idea was to investigate how web crawlers like Google and Yahoo! navigate a page by leading them into a maze of pages. The results of the experiment can still be viewed at *http://www.drunkmenworkhere.org/219*, though note that this experiment is pretty old — search engines have changed a lot in the past decade.

Take some time to play around with the page to see how it works.

Not Easy to Guess   You'll note that the pages here use a "r" URL parameter that is non-guessable. If this would not have been the case, that is,if all values for the URL parameter would have fallen between a well-defined range or look like successive numbers, writing a scraper would be much easier, as the list of URLs you want to obtain is then well-defined. Keep this in mind when exploring a page to see how viable it is to scrape it and to figure out which approach you'll have to take.

A first attempt to start scraping this site looks as follows:

```
import requests
from bs4 import BeautifulSoup
from urllib.parse import urljoin

base_url = 'http://www.webscrapingfordatascience.com/crawler/'
links_seen = set()

def visit(url, links_seen):
    html = requests.get(url).text
    html_soup = BeautifulSoup(html, 'html.parser')
    links_seen.add(url)
```

```
for link in html_soup.find_all("a"):
    link_url = link.get('href')
    if link_url is None:
        continue
    full_url = urljoin(url, link_url)
    if full_url in links_seen:
        continue
    print('Found a new page:', full_url)
    # Normally, we'd store the results here too
    visit(full_url, links_seen)

visit(base_url, links_seen)
```

Note that we're using the urljoin function here. The reason why we do so is because the "href" attribute of links on the page refers to a relative URL, for instance, "?r=f01e7f 02e91239a2003bdd35770e1173", which we need to convert to an absolute one. We could just do this by prepending the base URL, that is, base_url + link_url, but once we'd start to follow links and pages deeper in the site's URL tree, that approach would fail to work. Using urljoin is the proper way to take an existing absolute URL, join a relative URL, and get a well-formatted new absolute URL.

What About Absolute href Values?   The urljoin approach will even work with absolute "href" link values. Using:

```
urljoin('http://example.org', 'https://www.other.com/dir/')
```

will return "https://www.other.com/dir/", so this is always a good function to rely on when crawling.

If you run this script, you'll see that it will start visiting the different URLs. If you let it run for a while, however, this script will certainly crash, and not only because of network hiccups. The reason for this is because we use recursion: the visit function is calling itself over and over again, without an opportunity to go back up in the call tree as every page will contain links to other pages:

```
Traceback (most recent call last):
  File "C:\Users\Seppe\Desktop\firstexample.py", line 23, in <module>
```

```
    visit(url, links_seen)
[...]
    return wr in self.data
RecursionError: maximum recursion depth exceeded in comparison
```

As such, relying on recursion for web crawling is generally not a robust idea. We can rewrite our code as follows without recursion:

```python
import requests
from bs4 import BeautifulSoup
from urllib.parse import urljoin

links_todo = ['http://www.webscrapingfordatascience.com/crawler/']
links_seen = set()
def visit(url, links_seen):
    html = requests.get(url).text
    html_soup = BeautifulSoup(html, 'html.parser')
    new_links = []
    for link in html_soup.find_all("a"):
        link_url = link.get('href')
        if link_url is None:
            continue
        full_url = urljoin(url, link_url)
        if full_url in links_seen:
            continue
        # Normally, we'd store the results here too
        new_links.append(full_url)
    return new_links

while links_todo:
    url_to_visit = links_todo.pop()
    links_seen.add(url_to_visit)
    print('Now visiting:', url_to_visit)
    new_links = visit(url_to_visit, links_seen)
    print(len(new_links), 'new link(s) found')
    links_todo += new_links
```

You can let this code run for a while. This solution is better, but it still has several drawbacks. If our program crashes (e.g., when your Internet connection or the website is down), you'll have to restart from scratch again. Also, we have no idea how large the links_seen set might become. Normally, your computer will have plenty of memory available to easily store thousands of URLs, though we might wish to resort to a database to store intermediate progress information as well as the results.

6.3 Storing Results in a Database

Let's adapt our example to make it more robust against crashes by storing progress and result information in a database. We're going to use the "records" library to manage an SQLite database (a file based though powerful database system) in which we'll store our queue of links and retrieved numbers from the pages we crawl, which can be installed using pip:

```
pip install -U records
```

The adapted code, then, looks as follows:

```
import requests
import records
from bs4 import BeautifulSoup
from urllib.parse import urljoin
from sqlalchemy.exc import IntegrityError

db = records.Database('sqlite:///crawler_database.db')

db.query('''CREATE TABLE IF NOT EXISTS links (
        url text PRIMARY KEY,
        created_at datetime,
        visited_at datetime NULL)''')
db.query('''CREATE TABLE IF NOT EXISTS numbers (url text, number integer,
        PRIMARY KEY (url, number))''')

def store_link(url):
    try:
        db.query('''INSERT INTO links (url, created_at)
                VALUES (:url, CURRENT_TIMESTAMP)''', url=url)
```

```python
    except IntegrityError as ie:
        # This link already exists, do nothing
        pass

def store_number(url, number):
    try:
        db.query('''INSERT INTO numbers (url, number)
                    VALUES (:url, :number)''', url=url, number=number)
    except IntegrityError as ie:
        # This number already exists, do nothing
        pass

def mark_visited(url):
    db.query('''UPDATE links SET visited_at=CURRENT_TIMESTAMP
                WHERE url=:url''', url=url)

def get_random_unvisited_link():
    link = db.query('''SELECT * FROM links
                       WHERE visited_at IS NULL
                       ORDER BY RANDOM() LIMIT 1''').first()
    return None if link is None else link.url

def visit(url):
    html = requests.get(url).text
    html_soup = BeautifulSoup(html, 'html.parser')
    new_links = []
    for td in html_soup.find_all("td"):
        store_number(url, int(td.text.strip()))
    for link in html_soup.find_all("a"):
        link_url = link.get('href')
        if link_url is None:
            continue
        full_url = urljoin(url, link_url)
        new_links.append(full_url)
    return new_links
```

```
store_link('http://www.webscrapingfordatascience.com/crawler/')
url_to_visit = get_random_unvisited_link()
while url_to_visit is not None:
    print('Now visiting:', url_to_visit)
    new_links = visit(url_to_visit)
    print(len(new_links), 'new link(s) found')
    for link in new_links:
        store_link(link)
    mark_visited(url_to_visit)
    url_to_visit = get_random_unvisited_link()
```

From SQL to ORM   We're writing SQL (Structured Query Language) statements manually here to interact with the database, which is fine for smaller projects. For more complex projects, however, it's worthwhile to investigate some ORM (Object Relational Mapping) libraries thatare available in Python, such as SQLAlchemy or Pee-wee, which allow for a smoother and more controllable "mapping" between a relational database and Python objects, so that you can work with the latter directly, without having to deal with writing SQL statements. In the examples chapter, we'll use another library called "dataset" that also offers a convenient way to quickly dump information to a database without having to write SQL. Also note that you don't have to use the "records" library to use SQLite databases in Python. Python already comes with a "sqlite3" module built in (which is being used by records), which you could use as well (see *https://docs.python.org/3/ library/sqlite3.html*). The reason why we use records here is because it involves a bit less boilerplate code.

Try running this script for a while and then stop it (either by closing the Python window or pressing Control+C). You can take a look at the database ("crawler_database. db") using a SQLite client such as "DB Browser for SQLite," which can be obtained from *http://sqli tebrowser.org/*. Figure 6-1 shows this tool in action. Remember that you can also use the records library to fetch the stored results in your Python scripts as well.

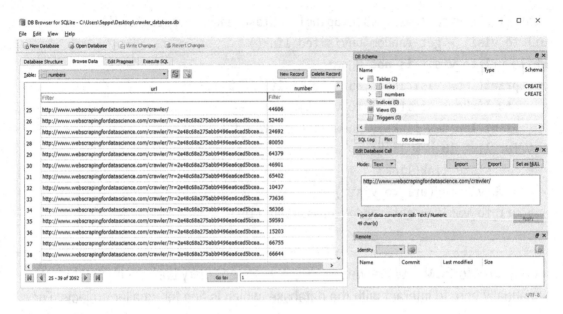

Figure 6-1. *Taking a look at the crawled results using DB Browser for SQLite*

SQLite   Note that the SQLite database we're using here is fine for smaller projects, but might give trouble once you start parallelizing your programs. Starting multiple instances of the script above will most likely work fine up to a certain degree, but might crash from time to time due to the SQLite database file being locked (i.e., used for too long by another script's process). Similarly, SQLite can be pretty daunting to work with when using multithreaded setups in your scripts as well. Switching to a server-client-oriented database like MySQL or Postgresql might be a good option in such cases.

Let's now try to use the same framework in order to build a crawler for Wikipedia. Our plan here is to store page titles, as well as keep track of "(from, to)" links on each page, starting from the main page. Note that our database scheme looks a bit different here:

```
import requests
import records
from bs4 import BeautifulSoup
from urllib.parse import urljoin, urldefrag
from sqlalchemy.exc import IntegrityError
```

```python
db = records.Database('sqlite:///wikipedia.db')

db.query('''CREATE TABLE IF NOT EXISTS pages (
            url text PRIMARY KEY,
            page_title text NULL,
            created_at datetime,
            visited_at datetime NULL)''')
db.query('''CREATE TABLE IF NOT EXISTS links (
            url text, url_to text,
            PRIMARY KEY (url, url_to))''')

base_url = 'https://en.wikipedia.org/wiki/'


def store_page(url):
    try:
        db.query('''INSERT INTO pages (url, created_at)
                    VALUES (:url, CURRENT_TIMESTAMP)''', url=url)
    except IntegrityError as ie:
        # This page already exists
        pass

def store_link(url, url_to):
    try:
        db.query('''INSERT INTO links (url, url_to)
                    VALUES (:url, :url_to)''', url=url, url_to=url_to)
    except IntegrityError as ie:
        # This link already exists
        pass

def set_visited(url):
    db.query('''UPDATE pages SET visited_at=CURRENT_TIMESTAMP
                WHERE url=:url''', url=url)

def set_title(url, page_title):
    db.query('UPDATE pages SET page_title=:page_title WHERE url=:url',
             url=url, page_title=page_title)
```

```python
def get_random_unvisited_page():
    link = db.query('''SELECT * FROM pages
                       WHERE visited_at IS NULL
                       ORDER BY RANDOM() LIMIT 1''').first()
    return None if link is None else link.url

def visit(url):
    print('Now visiting:', url)
    html = requests.get(url).text
    html_soup = BeautifulSoup(html, 'html.parser')
    page_title = html_soup.find(id='firstHeading')
    page_title = page_title.text if page_title else ''
    print(' page title:', page_title)
    set_title(url, page_title)
    for link in html_soup.find_all("a"):
        link_url = link.get('href')
        if link_url is None:
            # No href, skip
            continue
        full_url = urljoin(base_url, link_url)
        # Remove the fragment identifier part
        full_url = urldefrag(full_url)[0]
        if not full_url.startswith(base_url):
            # This is an external link, skip
            continue
        store_link(url, full_url)
        store_page(full_url)
    set_visited(url)

store_page(base_url)
url_to_visit = get_random_unvisited_page()
while url_to_visit is not None:
    visit(url_to_visit)
    url_to_visit = get_random_unvisited_page()
```

Note that we incorporate some extra measures to prevent visiting links we don't want to include: we make sure that we don't follow external links outside the base URL. We also use the `urldefrag` function to remove the fragment identifier (i.e., the part after "#") in URLs of links, as we don't want to visit the same page again even if it has a fragment identifier attached, as this will be equivalent to visiting and parsing the page without the fragment identifier. In other words, we don't want to include both "page.html#part1" and "page.html#part2" in our queue; simply including "page.html" is sufficient.

If you let this script run for a while, there are some fun things you can do with the collected data. For instance, you can try using the "NetworkX" library in Python to create a graph using the "(from, to)" links on each page. In the final chapter, we'll explore such use cases a bit further.

Note that we assume here that we know in advance which information we want to get out of the pages (the page title and links, in this case). In case you don't know beforehand what you'll want to get out of crawled pages, you might want to split up the crawling and parsing of pages even further, for example, by storing a complete copy of the HTML contents that can be parsed by a second script. This is the most versatile setup, but comes with additional complexity, as the following code shows:

```python
import requests
import records
import re
import os, os.path
from bs4 import BeautifulSoup
from urllib.parse import urljoin, urldefrag
from sqlalchemy.exc import IntegrityError

db = records.Database('sqlite:///wikipedia.db')

# This table keeps track of crawled and to-crawl pages
db.query('''CREATE TABLE IF NOT EXISTS pages (
          url text PRIMARY KEY,
          created_at datetime,
          html_file text NULL,
          visited_at datetime NULL)''')
```

```python
# This table keeps track of a-tags
db.query('''CREATE TABLE IF NOT EXISTS links (
            url text, link_url text,
            PRIMARY KEY (url, link_url))''')

# This table keeps track of img-tags
db.query('''CREATE TABLE IF NOT EXISTS images (
            url text, img_url text, img_file text,
            PRIMARY KEY (url, img_url))''')

base_url = 'https://en.wikipedia.org/wiki/'
file_store = './downloads/'

if not os.path.exists(file_store):
    os.makedirs(file_store)

def url_to_file_name(url):
    url = str(url).strip().replace(' ', '_')
    return re.sub(r'(?u)[^-\w.]', '', url)

def download(url, filename):
    r = requests.get(url, stream=True)
    with open(os.path.join(file_store, filename), 'wb') as the_image:
        for byte_chunk in r.iter_content(chunk_size=4096*4):
            the_image.write(byte_chunk)

def store_page(url):
    try:
        db.query('''INSERT INTO pages (url, created_at)
                    VALUES (:url, CURRENT_TIMESTAMP)''',
                 url=url)
    except IntegrityError as ie:
        pass

def store_link(url, link_url):
    try:
        db.query('''INSERT INTO links (url, link_url)
                    VALUES (:url, :link_url)''',
                 url=url, link_url=link_url)
```

```python
    except IntegrityError as ie:
        pass

def store_image(url, img_url, img_file):
    try:
        db.query('''INSERT INTO images (url, img_url, img_file)
                    VALUES (:url, :img_url, :img_file)''',
                url=url, img_url=img_url, img_file=img_file)
    except IntegrityError as ie:
        pass

def set_visited(url, html_file):
    db.query('''UPDATE pages
                SET visited_at=CURRENT_TIMESTAMP,
                    html_file=:html_file
                WHERE url=:url''',
            url=url, html_file=html_file)

def get_random_unvisited_page():
    link = db.query('''SELECT * FROM pages
                       WHERE visited_at IS NULL
                       ORDER BY RANDOM() LIMIT 1''').first()
    return None if link is None else link.url

def should_visit(link_url):
    link_url = urldefrag(link_url)[0]
    if not link_url.startswith(base_url):
        return None
    return link_url

def visit(url):
    print('Now visiting:', url)
    html = requests.get(url).text
    html_soup = BeautifulSoup(html, 'html.parser')
    # Store a-tag links
    for link in html_soup.find_all("a"):
        link_url = link.get('href')
```

```python
            if link_url is None:
                continue
            link_url = urljoin(base_url, link_url)
            store_link(url, link_url)
            full_url = should_visit(link_url)
            if full_url:
                # Queue for crawling
                store_page(full_url)
        # Store img-src files
        for img in html_soup.find_all("img"):
            img_url = img.get('src')
            if img_url is None:
                continue
            img_url = urljoin(base_url, img_url)
            filename = url_to_file_name(img_url)
            download(img_url, filename)
            store_image(url, img_url, filename)
        # Store the HTML contents
        filename = url_to_file_name(url) + '.html'
        fullname = os.path.join(file_store, filename)
        with open(fullname, 'w', encoding='utf-8') as the_html:
            the_html.write(html)
        set_visited(url, filename)

store_page(base_url)
url_to_visit = get_random_unvisited_page()
while url_to_visit is not None:
    visit(url_to_visit)
    url_to_visit = get_random_unvisited_page()
```

There is quite a lot going on here. First, three tables are created:

- A "pages" table to keep track of URLs, whether we already visited them
  or not (just like in the example before), but now also a "html_file" field
  referencing a file containing the HTML contents of the page;

- A "links" table to keep track of links found on pages;

- An "images" table to keep track of downloaded images.

For every page we visit, we extract all the "<a>" tags and store their links, and use a defined `should_visit` function to determine whether we should also queue the link in the "pages" table for crawling as well. Next, we inspect all "<img>" tags and download them to disk. A `url_to_file_name` function is defined in order to make sure that we get proper filenames using a regular expression. Finally, the HTML contents of the visited page are saved to disk as well, with the filename stored in the database. If you let this script run for a while, you'll end up with a large collection of HTML files and downloaded images, as Figure 6-2 shows.

Figure 6-2. *A collection of crawled images and web pages*

There are certainly many other modifications you could make to this setup. For instance, it might not be necessary to store links and images during the crawling of pages, since we could gather this information out from the saved HTML files as well. On the other hand, you might want to take a "save as much as possible as early as possible approach," and include other media elements as well in your download list. What should be clear is that web crawling scripts come in many forms and sizes, without there being a "one size fits all" to tackle every project. Hence, when working on a web crawler, carefully consider all design angles and how you're going to implement them in your setup.

Going Further   To mention some other modifications that are still possible to explore, you might, for instance, also wish to exclude "special" or "talk" pages, or pages that refer to a file from being crawled. Also keep in mind that our examples above do not account for the fact that information gathered from pages might become "stale." That is, once a page has been crawled, we'll never consider it again in our script. Try thinking about a way to change this (hint: the "visited_at" timestamp could be used to determine whether it's time to visit a page again, for instance). How would you balance exploring new pages versus refreshing older ones? Which set of pages would get priority, and how?

This concludes our journey through HTTP, requests, HTML, Beautiful Soup, JavaScript, Selenium, and crawling. The final chapter in this book contains some fully worked-out examples showing off some larger projects using real-life sites on the World Wide Web. First, however, we'll take a step back from the technicalities and discuss the managerial and legal aspects of web scraping.

PART III

Managerial Concerns and Best Practices

CHAPTER 7

Managerial and Legal Concerns

Up until now, we've been focusing a lot on the "web scraping" part of this book. We now take a step back and link the concepts you've learned to the general field of data science, paying particular attention to managerial issues that will arise when you're planning to incorporate web scraping in a data science project. This chapter also provides a thorough discussion regarding the legal aspects of web scraping.

7.1 The Data Science Process

As a data scientist (or someone aspiring to become one), you have probably already experienced that "data science" has become a very overloaded term indeed. Companies are coming to terms with the fact that data science incorporates a very broad skill set that is near impossible to find in a single person, hence the need for a multidisciplinary team involving:

- Fundamental theorists, mathematicians, statisticians (think: regression, Bayesian modeling, linear algebra, Singular Value Decomposition, and so on).

- Data wranglers (think: people who know their way around SQL, Python's pandas library and R's dplyr).

- Analysts and modelers (think: building a random forest or neural network model using R or SAS).

- Database administrators (think: DB2 or MSSQL experts, people with a solid understanding of databases and SQL).

© Seppe vanden Broucke and Bart Baesens 2018
S. vanden Broucke and B. Baesens, *Practical Web Scraping for Data Science*,
https://doi.org/10.1007/978-1-4842-3582-9_7

- Business intelligence experts (think: reporting and dashboards, as well as data warehouses and OLAP cubes).

- IT architects and "devops" people (think: people maintaining the modeling environment, tools, and platforms).

- Big data platform experts (think: people who know their way around Hadoop, Hive, and Spark).

- "Hacker" profiles (think: people comfortable on the command line, who know a bit of everything, can move fast, and break things).

- Business integrators, sponsors, and champions (think: people who know how to translate business questions to data requirements and modeling tasks, and can translate results back to stakeholders, and can emphasize the importance of data and data science in the organization).

- Management (think: higher ups who put a focus on data on the agenda and have it trickle down throughout all layers of the organization).

- And of course... web scrapers (think: what you've learned so far).

Data science and analytics continue to evolve at a rapid pace. The current hype surrounding AI and deep learning has given rise to yet again another collection of skills that organizations are looking for.

In any case, data science is about extracting value from data in a grounded manner where one realizes that data requires a lot of treatment and work from a lot of stakeholders before becoming valuable. As such, data science as an organizational activity is oftentimes described by means of a "process": a workflow describing the steps that have to be undertaken in a data science project, be it the construction of a predictive model to forecast who is going to churn, which customers will react positively to a marketing campaign, a customer segmentation task, or simply the automated creation of a periodical report listing some descriptive statistics. The raw source is data. Many such process frameworks have been proposed, with CRISP-DM and the KDD (Knowledge Discovery in Databases) process being the two most popular ones these days.

CRISP-DM stands for the "Cross-Industry Standard Process for Data Mining." Polls conducted by KDnuggets (*https://www.kdnuggets.com/*) over the past decade show that it is the leading methodology used by industry data miners and data scientists. Figure 7-1 shows a diagram depicting the CRISP-DM process.

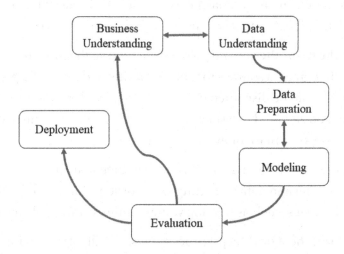

Figure 7-1. *The CRISP-DM process*

CRISP-DM is well-liked as it explicitly highlights the cyclic nature of data science: you'll often have to go back to the beginning to hunt for new data sources in case the outcome of a project is not in line with what you expected or had hoped for. The KDD process is a bit older than CRISP-DM and describes very similar steps (more as a straight-to path, though also here one has to keep in mind that going back a few steps can be necessary as well). It includes:

- **Identify the business problem:** Similar to the "Business Understanding" step in CRISP-DM, the first step consists of a thorough definition of the business problem. Some examples: customer segmentation of a mortgage portfolio, retention modeling for a postpaid telco subscription, or fraud detection for credit cards. Defining the scope of the analytical modeling exercise requires a close collaboration between the data scientist and business expert. Both need to agree on a set of key concepts such as: How do we define a customer, transaction, churn, fraud, etc.; what is it we want to predict (how do we define this), and when are we happy with the outcome.

- **Identify data sources:** Next, all source data that could be of potential interest needs to be identified. This is a very important step as data is the key ingredient to any analytical exercise and the selection of data has a deterministic impact on the analytical models built in a later step.

- **Select the data:** The general golden rule here is: the more data, the better, though data sources that have nothing to do with the problem at hand should be discarded during this step. All appropriate data will then be gathered in a staging area and consolidated into a data warehouse, data mart, or even a simple spreadsheet file.

- **Clean the data:** After the data has been gathered, a long preprocessing and data wrangling series of steps follows to remove all inconsistencies, such as missing values, outliers, and duplicate data.

- **Transform the data:** The preprocessing step will often also include a lengthy transformation part as well. Additional transformations may be considered, such as alphanumeric to numeric coding, geographical aggregation, logarithmic transformation to improve symmetry, and other smart "featurization" approaches.

- **Analyze the data:** The steps above correspond with the "Data Understanding" and "Data Preparation" steps in CRISP-DM. Once the data is sufficiently cleaned and processed, the actual analysis and modeling can begin (referred to as "Modeling" in CRISP-DM). Here, an analytical model is estimated on the preprocessed and transformed data. Depending upon the business problem, a particular analytical technique will be selected and implemented by the data scientist.

- **Interpret, evaluate, and deploy the model:** Finally, once the model has been built, it will be interpreted and evaluated by the business experts (denoted as "Evaluation" in CRISP-DM). Trivial patterns and insights that may be detected by the analytical model can still be interesting as they provide a validation of the model. But, of course, the key challenge is to find the unknown, yet interesting and actionable patterns that can provide new insights into your data. Once the analytical model has been appropriately validated and approved, it can be put into production as an analytics application (e.g., decision support system, scoring engine, etc.). It is important to consider how to represent the model output in a user-friendly way, how to integrate it with other applications (e.g., marketing campaign management tools, risk engines, etc.), and how to ensure the analytical model can be appropriately monitored on an ongoing basis. Often, the deployment of an analytical model will require the support of IT experts that will help to "productionize" the model.

7.2 Where Does Web Scraping Fit In?

There are various parts of the data science process where web scraping can fit in. In most projects, however, web scraping will form an important part in the identification and selection of data sources. That is, to collect and gather data you can include in your data set to be used for modeling.

It is important to provide a warning here, which is to always keep the production setting of your constructed models in mind (the "model train" versus "model run" gap). Are you building a model as a one-shot project that will be used to describe or find some interesting patterns, then by all means utilize as much scraped and external data as desired. In case a model will be productionized as a predictive analytics application, however, keep in mind that the model will need to have access to the same variables at the time of deployment as when it was trained. You'll hence carefully need to consider whether it will be feasible to incorporate scraped data sources in such a setup, as it needs to be ensured that the same sources will remain available and that it will be possible to continue scraping them as you go forward. Websites can change, and a data collection part depending on web scraping requires a great deal of maintenance to implement fixes or changes in a timely manner. In these cases, you still might wish to rely on a more robust solution like an API. Depending on your project, this requirement might be more or less troublesome to deal with. If the data you've scraped refers to aggregated data that "remains valid" for the duration of a whole year, for instance, then you can of course continue to use the collected data when running the models during deployment as well (and schedule a refresh of the data well before the year is over, for instance). Always keep the production setting of a model in mind: Will you have access to the data you need when applying and using the model as well, or only during the time of training the model? Who will be responsible to ensure this data access? Is the model simply a proof of concept with a limited shelf life, or will it be used and maintained for several years going forward?

In some cases, the web scraping part will form the main component of a data science project. This is common in cases where some basic statistics and perhaps an appealing visualization is built over scraped results to present findings and explore the gathered data in a user-friendly way. Still, the same questions have to be asked here: Is this a one-off report with a limited use time, or is this something people will want to keep up to date and use for a longer period of time?

The way how you answer these questions will have a great impact on the setup of your web scraper. In case you only need to gather results using web scraping for a quick proof of concept, a descriptive model, or a one-off report, you can afford to sacrifice robustness for the sake of obtaining data quickly. In case scraped data will be used during production as well (as was the case for the yearly aggregated information), it can still be feasible to scrape results, although it is a good idea to already think about the next time you'll have to refresh the data set and keep your setup as robust and well-documented as possible. If information has to be scraped every time the model is run, the "scraping" part now effectively becomes part of the deployed setup, including all headaches that come along with it regarding monitoring, maintenance, and error handling. Make sure to agree upfront which teams will be responsible for this!

There are two other "managerial warnings" we wish to provide here. One relates to data quality. If you've been working with data in an organizational setting, you've no doubt heard about the GIGO principle: garbage in, garbage out. When you rely on the World Wide Web to collect data — with all the messiness and unstructuredness that goes with it — be prepared to take a "hit" regarding data quality. Indeed, it is crucial to incorporate as much cleaning and fail-safes as possible in your scrapers, though you will nevertheless almost always eventually encounter a page where an extra unforeseen HTML tag appears or the text you expected is not there, or something is formatted just slightly differently. A final warning relates to reliability. The same point holds, in fact, not just for web scraping but also for APIs. Many promising startups over the past years have appeared that utilize Twitter's, Facebook's or some other API to provide a great service. What happens when the provide or owner of that website decides to increase their prices for what they're offering to others? What happens if they retire their offering? Many products have simply disappeared because their provider changed the rules. Using external data in general is oftentimes regarded as a silver bullet — "If only we could get out the information Facebook has!" — though think carefully and consider all possible outcomes before getting swayed too much by such ideas.

7.3 Legal Concerns

A few years ago, in 2015, a young American startup called hiQ Labs was featured by
The Economist, explaining their novel approach toward human resource analytics, an
application area of data science that was then rapidly getting increased attention, and is
still going strong today. The idea was to use vast amounts of data sets to help businesses
understand how much it would cost to lose a senior employee, predict who's going to
leave the firm, and who'd be the top-ranked candidates among potential new hires. In
August of 2017, it was revealed that most of the "vast data set" that hiQ Labs had and was
collecting originated from LinkedIn, owned by Microsoft.

LinkedIn and Microsoft executives, were understandably not so very happy with this
state of affairs. The data belonged to LinkedIn, so they thought, and a cease and desist
order was sent out to request hiQ to stop scraping LinkedIn's data, as well as various
technical measures were implemented to keep hiQ Labs' bots out.

The startup fired back with a lawsuit, demanding that LinkedIn should remove
the technical barriers it had established to prevent hiQ's use of data. LinkedIn tried to
argue that hiQ Labs violated the 1986 Computer Fraud and Abuse Act by scraping data
(CFAA is a piece of legislation that often gets brought up in similar cases). The judge
did not agree, raising concerns that LinkedIn was "unfairly leveraging its power in the
professional networking market for an anti-competitive purpose" (hiQ offered evidence
that LinkedIn was developing its own version of hiQ's talent-monitoring management
software), and even compared LinkedIn's argument to allowing website owners to "block
access by individuals or groups on the basis of race or gender discrimination." The court
explained that "a user does not access a computer without authorization by using bots,
even in the face of technical countermeasures, when the data it accesses is otherwise
open to the public."

The publicity around the case has certainly been positive for hiQ and has led to
more potential customers reaching out (see *https://www.bloomberg.com/news/featu
res/2017-11-15/the-brutal-fight-to-mine-your-data-and-sell-it-to-your-boss*).
While this story sounds promising for companies and individuals worried about the legal
aspects of web scraping, two important remarks have to be made here. First, the ruling
was made as a preliminary injunction, which — at the time of writing — might not be the
final outcome, as LinkedIn has announced that it will take its case to the 9th U.S. Circuit
Court of Appeals. Second, there is some subjectivity in "data being open to the public." In
this particular case, hiQ was allowed to scrape any data from LinkedIn profiles that could
be accessed without logging in to the service, that is, information that LinkedIn members

have designated as being publicly visible. It is not clear whether the same ruling would apply to pages that require users to log in. This allows for an easy loophole for data behemoths aiming to protect their data. Facebook, for instance, has been requiring users to log in to view its information, even for profiles that have indicated their information as being "public."

In 2014, Resultly, another startup providing a shopping app that constructs a catalog of items for sale by scraping various retailers, accidentally overloaded the servers of the TV retailer QVC, causing outages that, according to QVC, cost them two million dollars in revenue. Here too, QVC sought a preliminary injunction based on the Computer Fraud and Abuse Act. However, the court also ruled that "Resultly was not QVC's competitor, a disgruntled QVC employee, or an unhappy QVC customer aiming to cause damage to QVC's serve," and the startup hence lacked the required intent to damage QVC's system. The court also noted that QVC used Akamai's caching services, so Resultly's scraper accessed Akamai's servers, not QVC's.

Other court cases did not turn out so well for the scraping party:

- In the case of *Associated Press (AP) vs. Meltwater*, the media-monitoring firm Meltwater had been crawling AP websites and had extracted and republished significant amounts of text from AP news articles. Meltwater claimed that it was operating under fair use provisions of copyright laws, though the court found the case to be in favor of AP.

- In the case of *Ticketmaster vs. Riedel Marketing Group* (*RMG*), the latter was web scraping Ticketmaster's site so that it could harvest large quantities of desirable tickets for resale. Ticketmaster argued that RMG had agreed to the terms and conditions of the site but ignored them and the court held that RMG had infringed on Ticketmaster's copyrighted material.

- In *Craigslist vs. Naturemarket*, Naturemarket was scraping e-mail addresses from Craigslist's site. Craigslist sued, claiming copyright infringement, infringing on the Computer Fraud and Abuse Act, and breach of terms of use. The court awarded Craigslist a judgment of more than one million dollars.

- In *Ryanair Ltd vs. PR Aviation BV*, the European Court of Justice ruled that owners of publicly available databases, which do not fall under the protection of the Database Directive, are free to restrict the use of the data through contractual terms on their website. PR Aviation was extracting data from Ryanair's website in order to compare prices and book flights on payment of a commission. Ryanair required anyone accessing flight data on its website to tick a box to accept its terms and conditions, which included a prohibition on the automated extraction of data from their website for commercial purposes, without the airline's permission. The court found that Ryanair was free to create contractual limits on the use of its database, and the case was hence ruled in its favor.

- In 2006, Google got involved in a long-winding legal battle with Belgian media firm Copiepresse. A Prohibitory Injunction from the Court of First Instance of Belgium was sent, noting that Copiepresse had asked the court to order Google to withdraw all the articles, photographs, and graphics of its Belgian publications from the Google News site on the basis of copyright infringement concerns, and that the service also circumvented the advertising of the publishers who get a considerable amount of their revenue from these advertisements. On the basis of these findings, the court found the Copiepresse claim admissible and ordered Google to withdraw the Copiepresse content from all their sites. Google appealed the verdict, but, in retaliation, also stopped referencing for several months the sites of these newspapers on its main search engine (as the judgment mentioned "all Google's sites"). This led to an ugly battle between Copiepresse and Google, ending with the two of them reaching an agreement to include the sites again in 2011.

- In a related case, back in the United States, the Second Circuit of the Court of Appeal held that Google's book scanning of millions of books was in fact fair use even though the works fall under copyright protection, because of the transformative nature of its behavior under the fair use doctrine. The court also confirmed that facts are not protected by copyright, suggesting that harvesting factual data from a website is not in itself an infringement.

- In *Facebook vs. Power Ventures*, Facebook also claimed that the defendant has violated the CFAA and the CAN-SPAM Act, a federal law that prohibits sending commercial e-mails with materially misleading information. The judge ruled in favor of the plaintiff.

- In *LinkedIn vs. Doe Defendants*, LinkedIn (again) filed a suit against hundreds of unnamed individuals for using bots to harvest user profiles from its website. The case is still ongoing, and Scraping Hub (the company providing scraping services to the defendants) has been summoned to respond in court.

The list goes on. What is clear is that the legal landscape around web scraping is still evolving, and that many of the laws cited as being violated have not matured well in our digital age. In the United States, most court cases have involved one of the following theories of infringement or liability:

- **Breach of Terms and Conditions:** Most websites post terms and conditions or end-user license agreements on their pages, oftentimes explicitly addressing the issue of access to their website via scrapers. This is intended to create breach of contract liability by establishing a contract between the website owner and the scraper. However, posting such terms on a site may not be enough to show that a scraper has breached the terms of the website, since there is no active acceptance on the part of the scraper. What does appear to be more enforceable is the use of an explicit check box or "I accept" link in which the web scraper has to actively click to accept the terms. The same holds for scraping applications that log in to a site to access a nonpublic area, since the creation of an associated account will also often include explicit agreement of terms.

- **Copyright or Trademark Infringement:** In the United States, the legal doctrine of "fair use" allows limited use of copyrighted material under certain conditions without the explicit permission of the copyright holder. Uses for such purposes as parody, criticism, commentary, or academic research are regarded as fair use. Most commercial uses of copyrighted material do not fall under this doctrine, however.

- **Computer Fraud and Abuse Act (CFAA):** There are several federal and state laws against hacking or accessing another's computer. The CFAA states that whoever "intentionally accesses a computer without authorization... and as a result of such conduct recklessly causes damage" is basically in violation, especially if the violated website can prove loss or damage.

- **Trespass to Chattels:** This is a term referring to a civil wrong, which means one entity has interfered with another's personal property that causes loss of value or damage, and is also sometimes used as a theory of infringement in web scraping cases, like in a 1999 case of *eBay vs. Bidder's Edge*.

- **Robots Exclusion Protocol:** This is an industry standard that allows a website to embed a "robots.txt" file that communicates instructions to web crawlers to indicate which crawlers can access the site, and which pages they can access. It has limited legal value, however, but it might be a good idea to verify this file in your web scraping scripts to check whether the website owner wishes to keep crawlers and scrapers out.

- **The Digital Millennium Copyright Act (DMCA), CAN-SPAM Act:** These have also in some cases been included in web scraping court cases.

The situation in the European Union (EU) is governed by different legislation and legal systems, but many of the same principles apply, for example, relating to terms and conditions or copyrighted works. Most website and database owners in the EU have tended to rely on copyright infringement claims against screen scrapers. Some other key provisions are:

- **The EU Database Directive of 1996:** This directive provides legal protection for the creators of databases that are not covered by intellectual property rights, so that elements of a database that are not the author's own original creation are protected. In particular, it provides protection where there "has been qualitatively and/or quantitatively a substantial investment in either the obtaining, verification or presentation of the contents." The 2015 ruling by the European Court of Justice in a case concerning Ryanair greatly strengthened the ability of website operators to protect their content through contractual terms and conditions when they are not covered by the database directive.

- **The Computer Misuse Act and Trespass to Chattels:** In addition to intellectual property rights infringement, in theory website owners have other legal arguments against web scraping. As in the United States, in the United Kingdom a website owner could try to bring a common law tort claim for trespass to chattels. A website owner in the United Kingdom could also seek to rely on the Computer Misuse Act 1990, which prohibits unauthorized access to, or modification of, computer material.

It is clear to see that web scraping, especially on a large scale or for commercial reasons, comes with complicated legal implications. It is hence always advisable to consult a lawyer, appropriate experts, or a compliance officer before embarking on such projects, and to keep the following key principles in mind:

- **Get written permission:** The best way to avoid legal issues is to get written permission from a website's owner covering which data you can scrape and to what extent.

- **Check the terms of use:** These will often include explicit provisions against automated extraction of data. Oftentimes, a site's API will come with its own terms of use regarding usage, which you should check as well.

- **Public information only:** If a site exposes information publicly, without explicitly requiring acceptance of terms and conditions, moderated scraping is most likely fine. Sites that require you to log in is another story, however.

- **Don't cause damage:** Be nice when scraping! Don't hammer websites with lots of requests, overloading their network and blocking them of normal usage. Stay away from protected computers, and do not try to access servers you're not given access to.

- **Copyright and fair use:** Copyright law seems to provide the strongest means for plaintiffs to argue their case. Check carefully whether your scraping case would fall under fair use, and do not use copyrighted works in commercial projects.

CHAPTER 8

Closing Topics

You're now ready to get started with your own web scraping projects. This chapter wraps up by providing some closing topics. First, we provide an overview of other helpful tools and libraries you might wish to use in the context of web scraping, followed by a summary of best practices and tips to consider when web scraping.

8.1 Other Tools

8.1.1 Alternative Python Libraries

In this book, we've been working with Python 3 and the requests, Beautiful Soup, and Selenium libraries. Keep in mind that the Python ecosystem also provides a wealth of other libraries to handle HTTP messaging, like the built-in "urllib" module, from which we have used a few functions as well but can deal with all things HTTP, "httplib2" (see *https://github.com/httplib2/httplib2*), "urllib3" (see *https://urllib3.readthedocs.io/*), "grequests" (see *https://pypi.python.org/pypi/grequests*) and "aiohttp" (see *http://aiohttp.readthedocs.io/*).

   If you don't want to use Beautiful Soup, keep in mind that the Beautiful Soup library itself depends on an HTML parser to perform most of the bulk parsing work. It is hence also possible to use these lower-level parsers directly should you wish to do so. The "html.parser" module in Python provides such a parser, which we've been using already as the "engine" for Beautiful Soup, but it can be used directly as well, with an "lxml" and "html5lib" being popular alternatives. Some people eventually prefer this approach, as it can be argued that the additional overhead added by Beautiful Soup causes some slowdown. This is true, though we find that in most uses, you'll first have to deal with other issues before scraping speed becomes a real concern, like, for example, setting up a parallel scraping mechanism.

187

© Seppe vanden Broucke and Bart Baesens 2018
S. vanden Broucke and B. Baesens, *Practical Web Scraping for Data Science*,
https://doi.org/10.1007/978-1-4842-3582-9_8

8.1.2 Scrapy

There exist other noteworthy scraping libraries in the Python ecosystem than the ones we've mentioned so far. Scrapy (see *https://scrapy.org/*) is a comprehensive Python library for crawling websites and extracting structured data from websites, which is also quite popular. It deals with both the HTTP and HTML side of things, and provides a command-line tool to quickly set up, debug, and deploy web crawlers. Scrapy is a powerful tool that is also worth learning, even though its programming interface differs somewhat from requests and Beautiful Soup or Selenium — though this should not pose too many problems for you based on what you've learned. Especially in cases where you have to write a robust crawler, it can be useful to take a look at Scrapy as it provides many sane defaults for restarting scripts, crawling in parallel, data collection, and so on. The main advantage of using Scrapy is that it is very easy to deploy scrapers in "Scrapy Cloud" (see *https://scrapinghub.com/scrapy-cloud*), a cloud platform for running web crawlers. This is helpful in cases where you want to quickly run your scraper on a bunch of servers without hosting these yourself, though note that this service comes at a cost. An alternative would be to set up your scrapers on Amazon AWS or Google's Cloud Platform. A notable drawback of Scrapy is that is does not emulate a full browser stack. Dealing with JavaScript will hence be troublesome using this library. There does exist a plug-in that couples "Splash" (a JavaScript rendering service) with Scrapy, though this approach is a little cumbersome to set up and maintain.

8.1.3 Caching

Caching is another interesting aspect to discuss. We haven't spoken about caching a lot so far, though it is interesting to keep in mind that it is a good idea to implement some client-side caching solution while building your web scrapers, which will keep fetched web pages in its "memory." This avoids continuously hammering web servers with requests over and over again, which is especially helpful during development of your scripts (where you'll oftentimes restart a script to see if a bug has been fixed, you get the results you expect, and so on). A very interesting library to take a look at in this context is CacheControl (see *http://cachecontrol.readthedocs.io/en/latest/*), which can simply be installed through pip and directly used together with requests as follows:

```
import requests
from cachecontrol import CacheControl

session = requests.Session()
cached_session = CacheControl(session)
# You can now use cached_session like a normal session
# All GET requests will be cached
```

8.1.4 Proxy Servers

You can also take a look at setting up a local HTTP proxy server on your development machine. An HTTP proxy server acts as an intermediary for HTTP requests as follows: a client (a web browser or Python program) sends off an HTTP request, though now not by contacting the web server on the Internet, but sending it to the proxy server first. Depending on the configuration of the proxy server, it may decide to modify the request before sending it along to the true destination.

There are various reasons why the usage of an HTTP proxy server can come in handy. First, most proxy servers include options to inspect HTTP requests and replies, so that they offer a solid add-on on top of your browser's developer tools. Second, most proxy servers can be configured to enable caching, keeping the HTTP replies in their memory so that the end destination does not have to be contacted multiple times for subsequent similar requests. Finally, proxy servers are oftentimes used for anonymity reasons as well. Note that the destination web server will see an HTTP request coming in that originated from the HTTP proxy server, which does not necessarily need to run on your local development machine. As such, they're also used as a means to circumvent web scraping mitigation techniques where you might be blocked in case the web server sees too many requests coming from the same machine. In this case, you can either pay for a service providing a pool of HTTP proxy servers (see, e.g., *https://proxymesh.com/*) or by using anonymity services such as Tor (see *https://www.torproject.org/*), which is free, but primarily meant to provide anonymity, and might be less suitable for web scrapers, as Tor tends to be relatively slow, and many web scraping mitigations will also keep a list of Tor "exit points" and block them. For some solid HTTP proxy server implementations, take a look at Squid (*http://www.squid-cache.org/*) or Fiddler (*https://www.telerik.com/fiddler*).

8.1.5 Scraping in Other Programming Languages

Moving away from Python to other languages, it is good to know that Selenium also provides libraries for a number of other programming languages, with Java being one of the other main targets of the project. In case you're working with R — another popular language for data science — definitely take a look at the "rvest" library (see *https://cran.r-project.org/web/packages/rvest/*), which is inspired by libraries like Beautiful Soup and makes it easy to scrape data from HTML pages from R. The recent rise in attention being given to JavaScript and the language becoming more and more viable as a server-side scripting language as well has also spawned a lot of powerful scraping libraries for this language as well. PhantomJS has become a very popular choice here (see *http://phantomjs.org/*), which emulates a full "headless" browser (and can be used as a "driver" in Selenium as well). Since PhantomJS code can be a bit verbose, other libraries such as Nightmare (see *http://www.nightmarejs.org/*) have been proposed, which offer a more user-friendly, high-level API on top of PhantomJS. Other interesting projects in this space are SlimerJS (see *https://slimerjs.org/*) that is similar to PhantomJS, except that it runs on top of Gecko, the browser engine of Mozilla Firefox, in-stead of Webkit. CasperJS is another high-level library (see *http://casperjs.org/*) that can be used on top of either PhantomJS or SlimerJS. Another interesting recent project is Puppeteer (see *https://github.com/GoogleChrome/puppeteer*), a library that provides a high-level API to control a headless Chrome web browser. Driven by the popularity of PhantomJS, Chrome's developers are spending a lot of effort in providing a headless version their browser. Up until now, most deployments of web scrapers relying on a full browser would either use PhantomJS, which is already headless but differs slightly from a real browser; or would use a Firefox or Chrome driver together with a virtual display, using "Xvfb" on Linux, for instance. Now that a true headless version of Chrome is becoming more stable, it is becoming a strong contender for PhantomJS, especially as it is also possible to use this headless Chrome setup as a driver for Selenium. Vitaly Slobodin, the former maintainer of PhantomJS, already stated that "people will switch to it, eventually. Chrome is faster and more stable than PhantomJS. And it doesn't eat memory like crazy." In March 2018, the maintainers of PhantomJS announced that the project would cease to receive further updates, urging users to switch to Puppeteer in- stead (see *https://github.com/ariya/phantomjs/issues/15344*). Using Puppeteer together with Python and Selenium is still somewhat "bleeding edge", but does work. Take a look at *https://intoli.com/blog/running-selenium-with-headless-chrome/* if you're interested in this approach.

8.1.6 Command-Line Tools

It's also worth mentioning a number of helpful command-line tools that can come in handy when debugging web scrapers and interactions with HTTP servers. HTTPie (see *https://httpie.org/*) is a command-line tool with beautiful output and support for form data, sessions, and JSON, making the tool also very helpful when debugging web APIs.

"cURL" is another older though very feature-rich and robust command-line tool, supporting much more than only HTTP (see *https://curl.haxx.se/*). This tool works particularly well with Chrome's Developer Tools, as you can right-click any request in the "Network" tab, and pick "Copy as cURL." This will put a command on your clipboard that you can paste in a command-line window, which will perform the exact same request as Chrome did, and should provide you with the same result. If you are stuck in a difficult debugging session, inspecting this command might give an indication about which header or cookie you're not including in your Python script.

8.1.7 Graphical Scraping Tools

Finally, we also need to talk a little bit about graphical scraping tools. These can be offered either offered as stand-alone programs or through a browser plug-in. Some of these have a free offering, like Portia (*https://portia.scrapinghub.com*) or Parsehub (*https://www.parsehub.com*) whereas others, like Kapow (https://www.kofax.com/data-integration-extraction), Fminer (http://www.fminer.com/), and Dexi (https://dexi.io/) are commercial offerings. The feature set of such tools differs somewhat. Some will focus on the user-friendliness aspect of their offering. "Just point us to a URL and we'll get out the interesting data," as if done by magic. This is fine, but based on what you've seen so far, you'll be able to replicate the same behavior in a much more robust (and possibly even quicker) way. As we've seen, getting out tables and a list of links is easy. Oftentimes, these tools will fail to work once the data contained in a page is structured in a less-straightforward way or when the page relies on JavaScript. Higher-priced offerings will be somewhat more robust. These tools will typically also emulate a full browser instance and will allow you to write a scraper using a workflow-oriented approach by dragging and dropping and configuring various steps. Figure 8-1 shows a screenshot of Kapow, which takes such an approach. This is already better, though it also comes with a number of drawbacks. First, many of these offerings are quite expensive, making them less viable for experiments, proof of concepts, or smaller projects. Second, elements will often be retrieved from a page based on a user simply clicking on the item they wish to retrieve. What happens in the background is that a CSS selector or XPath rule will be constructed, matching the selected element. This is fine in itself (we've also done this in our scripts), though note that a program is not as smart as a human programmer in terms of fine-tuning this rule. In many cases, a very granular, specific rule is constructed, which will break once the site returns results that are structured a bit differently or an update is made to the web page. Just as with scrapers written in Python, you'll hence have to keep in mind that you'll still need to maintain your collection of scrapers in case you wish to use them over longer periods of time. Graphical tools do not fix this issue for you, and can even cause scrapers to fail more quickly as the underlying "rules" that are constructed can be very specific. Many tools will allow you to change the selector rules manually, but to do so, you will have to learn how they work, and taking a programming-oriented approach might then quickly become more appealing anyway. Finally, in our experience with these tools, we've also noted that the browser stacks they include are not always that robust or recent. We've seen various cases where the internal browser simply crashes when confronted with a JavaScript-heavy page.

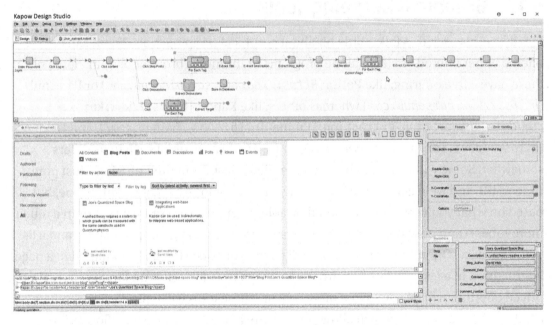

Figure 8-1. *An example of a graphical web scraping tool*

8.2 Best Practices and Tips

At various occasions, we've provided tips throughout this book. With the following overview, we provide a list of best practices to summarize what you should keep in mind when building web scrapers:

- **Go for an API first:** Always check first whether the site you wish to scrape offers an API. If it doesn't, or it doesn't provide the information you want, or it applies rate limiting, then you can decide to go for a web scraper instead.

- **Don't parse HTML manually:** Use a parser such as Beautiful Soup instead of trying to untangle the soup manually or using regular expressions.

- **Play nice:** Don't hammer a website with hundreds of HTTP requests, as this will end up with a high chance of you getting blocked anyway. Consider contacting the webmaster of the site and work out a way to work together.

- **Consider the user agent and referrer:** Remember the "User-Agent" and "Referer" headers. Many sites will check these to prevent scraping or unauthorized access.

- **Web servers are picky:** Whether it's URL parameters, headers, or form data, some web servers come with very picky and strange requirements regarding their ordering, presence, and values. Some might even deviate from the HTTP standard.

- **Check your browser:** If you can't figure out what's going wrong, start from a fresh browser session and use your browser's developer tools to follow along through a normal web session — preferably opened as an "Incognito" or "private browsing" window (to make sure you start from an empty set of cookies). If everything goes well there, you should be able to simulate the same behavior as well. Remember that you can use "curl" and other command-line tools to debug difficult cases.

- **Before going for a full JavaScript engine, consider internal APIs:** Check your browser's network requests to see whether you can access a data source used by JavaScript directly before going for a more advanced solution like Selenium.

- **Assume it will crash:** The web is a dynamic place. Make sure to write your scrapers in such a way that they provide early and detailed warnings when something goes wrong.

- **Crawling is hard:** When writing an advanced crawler, you'll quickly need to incorporate a database, deal with restarting scripts, monitoring, queue management, timestamps, and so on to create a robust crawler.

- **What about CAPTCHA's:** If a page shows a CAPTCHA (Completely Automated Public Turing test to tell Computers and Humans Apart), this is basically a way to announce that no scrapers are welcome. There are ways around this, however. Some sites offer a "CAPTCHA solving API" (with real humans behind it, see, for example, *http://www.deathbycaptcha.com*), offering quick solving times for a fair price. Some real-life projects have used OCR software such as Tesseract (see *https://github.com/tesseract-ocr* and *http://resources.infosecinstitute.com/case-study-cracking-online-banking-captcha-login-using-python/* for an example use case) to build a CAPTCHA solver instead. You will also find deep learning implementations to solve CAPTCHA's, for example, using convolutional neural networks (see, e.g., *https://medium.com/towards-data-science/deep-learning-drops-breaking-captcha-20c8fc96e6a3*, *https://medium.com/@ageitgey/how-to-break-a-captcha-system-in-15-minutes-with-machine-learning-dbebb035a710* or *http://www.npr.org/sections/thetwo-way/2017/10/26/560082659/ai-model-fundamentally-cracks-captchas-scientists-say* for recent stories on this topic). It's also a good idea to verify whether the CAPTCHA only appears after a number of requests or randomly, in which case you can implement a backing-off or cooldown strategy to simply wait for a while before trying again. Providing an exhaustive overview on how to break CAPTCHA's using automated approaches is out of scope for this work, though the final example in the next chapter provides some pointers to get you started on breaking them using deep learning.

- **Some tools are helpful, some are not:** There are various companies offering "cloud scraping" solutions, like, for example, Scrapy. The main benefit of using these is that you can utilize their fleet of servers to quickly parallelize a scraper. Don't put too much trust in expensive graphical scraping tools, however. In most cases, they'll only work with basic pages, cannot deal with JavaScript, or will lead to the construction of a scraping pipeline that might work but uses very fine-grained and specific selector rules that will break the moment the site changes its HTML a little bit.

- **Scraping is a cat-and-mouse game:** Some websites go very far in order to prevent scraping. Some researchers have investigated the various ways in which, for example, Selenium or browsers such as PhantomJS differ from normal ones (by inspecting their headers or JavaScript capabilities). It's possible to work around these checks, but there will be a point where it will become very hard to scrape a particular site. See, for instance, *https://blog. shapesecurity.com/2015/01/22/detecting-phantomjs-based-visitors/* for an interesting overview on detecting PhantomJS-based visitors or *https://intoli.com/blog/making-chrome-headless-undetectable/* for a similar overview regarding Headless Chrome. Even when using Chrome through Selenium, specialized solutions exist that will try to identify nonhuman patterns, such as scrolling or navigating or clicking too fast, always clicking at the middle position of an element, and so on. It's unlikely that you'll encounter many such cases in your projects, but keep this in mind nonetheless.

- **Keep in mind the managerial and legal concerns, and where web scraping fits in your data science process:** As discussed, consider the data quality, robustness, and deployment challenges that come with web scraping. Similarly, keep in mind the potential legal issues that might arise when you start depending on web scraping a lot or start to misuse it.

CHAPTER 9

Examples

This chapter includes several larger examples of web scrapers. Contrary to most of the examples showcased during the previous chapters, the examples here serve a twofold purpose. First, they showcase some more examples using real-life websites instead of a curated, safe environment. The reason why we haven't used many real-life examples so far is due to the dynamic nature of the web. It might be that the examples covered here do not provide the exact same results anymore or will be broken by the time you read them. That being said, we have tried to use a selection of sites that are rather scraper friendly and not very prone to changes. The second purpose of these examples is to highlight how various concepts seen throughout the book "fall together" and interact, as well as to hint toward some interesting data science-oriented use cases.

The following examples are included in this chapter:

- **Scraping Hacker News:** This example uses requests and Beautiful Soup to scrape the Hacker News front page.

- **Using the Hacker News API:** This example provides an alternative by showing how you can use APIs with requests.

- **Quotes to Scrape:** This example uses requests and Beautiful Soup and introduces the "dataset" library as an easy means to store data.

- **Books to Scrape:** This example uses requests and Beautiful Soup, as well as the dataset library, illustrating how you can run a scraper again without storing duplicate results.

- **Scraping GitHub Stars:** This example uses requests and Beautiful Soup to scrape GitHub repositories and show how you can perform a login using requests, reiterating our warnings regarding legal concerns.

© Seppe vanden Broucke and Bart Baesens 2018
S. vanden Broucke and B. Baesens, *Practical Web Scraping for Data Science*,
https://doi.org/10.1007/978-1-4842-3582-9_9

- **Scraping Mortgage Rates:** This example uses requests to scrape mortgage rates using a particularly tricky site.

- **Scraping and Visualizing IMDB Ratings:** This example uses requests and Beautiful Soup to get a list of IMDB ratings for TV series episodes. We also introduce the "matplotlib" library to create plots in Python.

- **Scraping IATA Airline Information:** This example uses requests and Beautiful Soup to scrape airline information from a site that employs a difficult web form. An alternative approach using Selenium is also provided. Scraped results are converted to a tabular format using the "pandas" library, also introduced in this example.

- **Scraping and Analyzing Web Forum Interactions:** This example uses requests and Beautiful Soup to scrape web forum posts and stores them using the dataset library. From the collected results, we use pandas and matplotlib to create heat map plots showing user activity.

- **Collecting and Clustering a Fashion Data Set:** This example uses requests and Beautiful Soup to download a set of fashion images. The images are then clustered using the "scikit-learn" library.

- **Sentiment Analysis of Scraped Amazon Reviews:** This example uses requests and Beautiful Soup to scrape a list of user reviews from Amazon, stored using the dataset library. We then analyze these using the "nltk" and "vaderSentiment" libraries in Python, and plot the results using matplotlib.

- **Scraping and Analyzing News Articles:** This example uses Selenium to scrape a list of news articles, stored using the dataset library. We then associate these to a list of topics by constructing a topic model using nltk.

- **Scraping and Analyzing a Wikipedia Graph:** In this example, we extended our Wikipedia crawler to scrape pages using requests and Beautiful Soup, stored using the dataset library, which we then use to create a graph using "NetworkX" and visualize it with matplotlib.

- **Scraping and Visualizing a Board Members Graph:** This example uses requests and Beautiful Soup to scrape board members for S&P 500 companies. A graph is created using NetworkX and visualized using "Gephi."

- **Breaking CAPTCHA's Using Deep Learning:** This example shows how a convolutional neural network can be used to break CAPTCHA's.

Source Code   The source code for all examples is also provided at the companion website for this book at *http://www.webscrapingfordatascience.com*.

9.1 Scraping Hacker News

We're going to scrape the *https://news.ycombinator.com/news* front page, using requests and Beautiful Soup. Take some time to explore the page if you haven't heard about it already. Hacker News is a popular aggregator of news articles that "hackers" (computer scientists, entrepreneurs, data scientists) find interesting.

We'll store the scraped information in a simple Python list of dictionary objects for this example. The code to scrape this page looks as follows:

```
import requests
import re
from bs4 import BeautifulSoup

articles = []

url = 'https://news.ycombinator.com/news'

r = requests.get(url)
html_soup = BeautifulSoup(r.text, 'html.parser')

for item in html_soup.find_all('tr', class_='athing'):
    item_a = item.find('a', class_='storylink')
    item_link = item_a.get('href') if item_a else None
```

199

```
    item_text = item_a.get_text(strip=True) if item_a else None
    next_row = item.find_next_sibling('tr')
    item_score = next_row.find('span', class_='score')
    item_score = item_score.get_text(strip=True) if item_score else '0 points'
    # We use regex here to find the correct element
    item_comments = next_row.find('a', string=re.compile('\d+( |\s)
    comment(s?)'))
    item_comments = item_comments.get_text(strip=True).replace('\xa0', ' ') \
                        if item_comments else '0 comments'
    articles.append({
        'link' : item_link,
        'title' : item_text,
        'score' : item_score,
        'comments' : item_comments})

for article in articles:
    print(article)
```

This will output the following:

```
{'link': 'http://moolenaar.net/habits.html', 'title': 'Seven habits of    ↵
    effective text editing (2000)', 'score': '44 points', 'comments':    ↵
    '9 comments'}
{'link': 'https://www.repository.cam.ac.uk/handle/1810/251038', 'title':    ↵
    'Properties of expanding universes (1966)', 'score': '52 points',    ↵
    'comments': '8 comments'}
[...]
```

Try expanding this code to scrape a link to the comments page as well. Think about potential use cases that would be possible when you also scrape the comments themselves (for example, in the context of text mining).

9.2 Using the Hacker News API

Note that Hacker News also offers an API providing structured, JSON-formatted results (see *https://github.com/HackerNews/API*). Let's rework our Python code to now serve as an API client without relying on Beautiful Soup for HTML parsing:

```python
import requests

articles = []

url = 'https://hacker-news.firebaseio.com/v0'

top_stories = requests.get(url + '/topstories.json').json()

for story_id in top_stories:
    story_url = url + '/item/{}.json'.format(story_id)
    print('Fetching:', story_url)
    r = requests.get(story_url)
    story_dict = r.json()
    articles.append(story_dict)

for article in articles:
    print(article)
```

This will output the following:

```
Fetching: https://hacker-news.firebaseio.com/v0/item/15532457.json
Fetching: https://hacker-news.firebaseio.com/v0/item/15531973.json
Fetching: https://hacker-news.firebaseio.com/v0/item/15532049.json
[...]
{'by': 'laktak', 'descendants': 30, 'id': 15532457, 'kids': [15532761, ↵
    15532768, 15532635, 15532727, 15532776, 15532626, 15532700, 15532634], ↵
    'score': 60, 'time': 1508759764, 'title': 'Seven habits of effective ↵
    text editing (2000)', 'type': 'story', 'url': 'http://moolenaar.net/ ↵
    habits.html'}
[...]
```

9.3 Quotes to Scrape

We're going to scrape *http://quotes.toscrape.com*, using requests and Beautiful Soup. This page is provided by Scrapinghub as a more realistic scraping playground. Take some time to explore the page. We'll scrape out all the information, that is:

- The quotes, with their author and tags;

- And the author information, that is, date and place of birth, and description.

We'll store this information in a SQLite database. Instead of using the "records" library and writing manual SQL statements, we're going to use the "dataset" library (see *https://dataset.readthedocs.io/en/latest/*). This library provides a simple abstraction layer removing most direct SQL statements without the necessity for a full ORM model, so that we can use a database just like we would with a CSV or JSON file to quickly store some information. Installing a dataset can be done easily through pip:

```
pip install -U dataset
```

Not a Full ORM   Note that dataset does not want to replace a full-blown ORM (Object Relational Mapping) library like SQLAlchemy (even though it uses SQLAlchemy behind the scenes). It's meant simply to quickly store a bunch of data in a database without having to define a schema or write SQL. For more advanced use cases, it's a good idea to consider using a true ORM library or to define a database schema by hand and query it manually.

The code to scrape this site looks as follows:

```python
import requests
import dataset
from bs4 import BeautifulSoup
from urllib.parse import urljoin, urlparse

db = dataset.connect('sqlite:///quotes.db')

authors_seen = set()
```

```
base_url = 'http://quotes.toscrape.com/'

def clean_url(url):
    # Clean '/author/Steve-Martin' to 'Steve-Martin'
    # Use urljoin to make an absolute URL
    url = urljoin(base_url, url)
    # Use urlparse to get out the path part
    path = urlparse(url).path
    # Now split the path by '/' and get the second part
    # E.g. '/author/Steve-Martin' -> ['', 'author', 'Steve-Martin']
    return path.split('/')[2]

def scrape_quotes(html_soup):
    for quote in html_soup.select('div.quote'):
        quote_text = quote.find(class_='text').get_text(strip=True)
        quote_author_url = clean_url(quote.find(class_='author') \
                                    .find_next_sibling('a').get('href'))
        quote_tag_urls = [clean_url(a.get('href'))
                          for a in quote.find_all('a', class_='tag')]
        authors_seen.add(quote_author_url)
        # Store this quote and its tags
        quote_id = db['quotes'].insert({ 'text' : quote_text,
                                         'author' : quote_author_url })
        db['quote_tags'].insert_many(
                [{'quote_id' : quote_id, 'tag_id' : tag} for tag in
                quote_tag_urls])

def scrape_author(html_soup, author_id):
    author_name = html_soup.find(class_='author-title').get_text(strip=True)
    author_born_date = html_soup.find(class_='author-born-date').get_text
    (strip=True)
    author_born_loc = html_soup.find(class_='author-born-location').
    get_text(strip=True)
    author_desc = html_soup.find(class_='author-description').get_text
    (strip=True)
```

```
db['authors'].insert({ 'author_id' : author_id,
                       'name' : author_name,
                       'born_date' : author_born_date,
                       'born_location' : author_born_loc,
                       'description' : author_desc})

# Start by scraping all the quote pages
url = base_url
while True:
    print('Now scraping page:', url)
    r = requests.get(url)
    html_soup = BeautifulSoup(r.text, 'html.parser')
    # Scrape the quotes
    scrape_quotes(html_soup)
    # Is there a next page?
    next_a = html_soup.select('li.next > a')
    if not next_a or not next_a[0].get('href'):
        break
    url = urljoin(url, next_a[0].get('href'))

# Now fetch out the author information
for author_id in authors_seen:
    url = urljoin(base_url, '/author/' + author_id)
    print('Now scraping author:', url)
    r = requests.get(url)
    html_soup = BeautifulSoup(r.text, 'html.parser')
    # Scrape the author information
    scrape_author(html_soup, author_id)
```

This will output the following:

```
Now scraping page: http://quotes.toscrape.com/
Now scraping page: http://quotes.toscrape.com/page/2/
Now scraping page: http://quotes.toscrape.com/page/3/
Now scraping page: http://quotes.toscrape.com/page/4/
Now scraping page: http://quotes.toscrape.com/page/5/
Now scraping page: http://quotes.toscrape.com/page/6/
```

```
Now scraping page: http://quotes.toscrape.com/page/7/
Now scraping page: http://quotes.toscrape.com/page/8/
Now scraping page: http://quotes.toscrape.com/page/9/
Now scraping page: http://quotes.toscrape.com/page/10/
Now scraping author: http://quotes.toscrape.com/author/Ayn-Rand
Now scraping author: http://quotes.toscrape.com/author/E-E-Cummings
[...]
```

Note that there are still a number of ways to make this code more robust. We're not checking for None results when scraping the quote or author pages. In addition, we're using "dataset" here to simply insert rows in three tables. In this case, dataset will automatically increment a primary "id" key. If you want to run this script again, you'll hence first have to clean up the database to start fresh, or modify the script to allow for resuming its work or updating the results properly. In later examples, we'll use dataset's upsert method to do so.

Once the script has finished, you can take a look at the database ("quotes.db") using a SQLite client such as "DB Browser for SQLite," which can be obtained from *http:// sqlitebrowser.org/*. Figure 9-1 shows this tool in action.

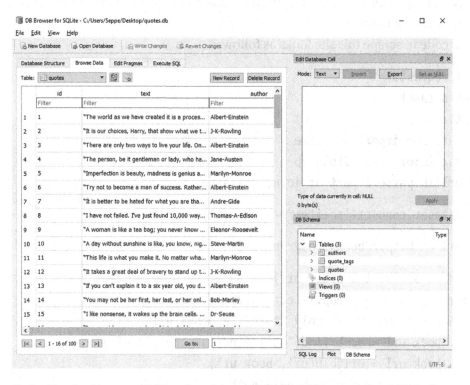

Figure 9-1. *Exploring an SQLite database with "DB Browser for SQLite"*

9.4 Books to Scrape

We're going to scrape *http://books.toscrape.com*, using requests and Beautiful Soup. This page is provided by Scrapinghub as a more realistic scraping playground. Take some time to explore the page. We'll scrape out all the information, that is, for every book, we'll obtain:

- Its title;

- Its image;

- Its price and stock availability;

- Its rating;

- Its product description;

- Other product information.

We're going to store this information in an SQLite database, again using the "dataset" library. However, this time we're going to write our program in such a way that it takes into account updates — so that we can run it multiple times without inserting duplicate records in the database.

The code to scrape this site looks as follows:

```python
import requests
import dataset
import re
from datetime import datetime
from bs4 import BeautifulSoup
from urllib.parse import urljoin, urlparse

db = dataset.connect('sqlite:///books.db')

base_url = 'http://books.toscrape.com/'

def scrape_books(html_soup, url):
    for book in html_soup.select('article.product_pod'):
        # For now, we'll only store the books url
        book_url = book.find('h3').find('a').get('href')
        book_url = urljoin(url, book_url)
        path = urlparse(book_url).path
```

```
        book_id = path.split('/')[2]
        # Upsert tries to update first and then insert instead
        db['books'].upsert({'book_id' : book_id,
                            'last_seen' : datetime.now()
                            }, ['book_id'])

def scrape_book(html_soup, book_id):
    main = html_soup.find(class_='product_main')
    book = {}
    book['book_id'] = book_id
    book['title'] = main.find('h1').get_text(strip=True)
    book['price'] = main.find(class_='price_color').get_text(strip=True)
    book['stock'] = main.find(class_='availability').get_text(strip=True)
    book['rating'] = ' '.join(main.find(class_='star-rating') \
                        .get('class')).replace('star-rating', '').strip()
    book['img'] = html_soup.find(class_='thumbnail').find('img').get('src')
    desc = html_soup.find(id='product_description')
    book['description'] = ''
    if desc:
        book['description'] = desc.find_next_sibling('p') \
                            .get_text(strip=True)
    info_table = html_soup.find(string='Product Information').find_
    next('table')
    for row in info_table.find_all('tr'):
        header = row.find('th').get_text(strip=True)
        # Since we'll use the header as a column, clean it a bit
        # to make sure SQLite will accept it
        header = re.sub('[^a-zA-Z]+', '_', header)
        value = row.find('td').get_text(strip=True)
        book[header] = value
    db['book_info'].upsert(book, ['book_id'])

# Scrape the pages in the catalogue
url = base_url
inp = input('Do you wish to re-scrape the catalogue (y/n)? ')
```

```python
while True and inp == 'y':
    print('Now scraping page:', url)
    r = requests.get(url)
    html_soup = BeautifulSoup(r.text, 'html.parser')
    scrape_books(html_soup, url)
    # Is there a next page?
    next_a = html_soup.select('li.next > a')
    if not next_a or not next_a[0].get('href'):
        break
    url = urljoin(url, next_a[0].get('href'))

# Now scrape book by book, oldest first
books = db['books'].find(order_by=['last_seen'])
for book in books:
    book_id = book['book_id']
    book_url = base_url + 'catalogue/{}'.format(book_id)
    print('Now scraping book:', book_url)
    r = requests.get(book_url)
    r.encoding = 'utf-8'
    html_soup = BeautifulSoup(r.text, 'html.parser')
    scrape_book(html_soup, book_id)
    # Update the last seen timestamp
    db['books'].upsert({'book_id' : book_id,
                        'last_seen' : datetime.now()
                        }, ['book_id'])
```

Once the script has finished, remember that you can take a look at the database ("books.db") using, for example, "DB Browser for SQLite." Note the use of the dataset's upsert method in this example. This method will try to update a record if it exists already (by matching existing records with a list of given field names), or insert a new record otherwise.

9.5 Scraping GitHub Stars

We're going to scrape *https://github.com*, using requests and Beautiful Soup. Our goal is to get, for a given GitHub username, like, for example, *https://github.com/google*, a list of repositories with their GitHub-assigned programming language as well as the number of stars a repository has.

The basic structure of this scraper is quite simple:

```python
import requests
from bs4 import BeautifulSoup
import re

session = requests.Session()

url = 'https://github.com/{}'
username = 'google'

r = session.get(url.format(username), params={'page': 1, 'tab':
'repositories'})
html_soup = BeautifulSoup(r.text, 'html.parser')
repos = html_soup.find(class_='repo-list').find_all('li')
for repo in repos:
    name = repo.find('h3').find('a').get_text(strip=True)
    language = repo.find(attrs={'itemprop': 'programmingLanguage'})
    language = language.get_text(strip=True) if language else 'unknown'
    stars = repo.find('a', attrs={'href': re.compile('\/stargazers')})
    stars = int(stars.get_text(strip=True).replace(',', '')) if stars else 0
    print(name, language, stars)
```

Running this will output:

```
sagetv Java 192
ggrc-core Python 233
gapid Go 445
certificate-transparency-rfcs Python 55
mtail Go 936
[...]
```

However, this will fail if we would try to scrape a normal user's page. Google's GitHub account is an enterprise account, which is displayed slightly differently from normal user accounts. You can try this out by setting the "username" variable to "Macuyiko" (one of the authors of this book). We hence need to adjust our code to handle both cases:

```python
import requests
from bs4 import BeautifulSoup
import re

session = requests.Session()

url = 'https://github.com/{}'
username = 'Macuyiko'

r = session.get(url.format(username), params={'page': 1, 'tab':
'repositories'})
html_soup = BeautifulSoup(r.text, 'html.parser')

is_normal_user = False
repos_element = html_soup.find(class_='repo-list')
if not repos_element:
    is_normal_user = True
    repos_element = html_soup.find(id='user-repositories-list')

repos = repos_element.find_all('li')
for repo in repos:
    name = repo.find('h3').find('a').get_text(strip=True)
    language = repo.find(attrs={'itemprop': 'programmingLanguage'})
    language = language.get_text(strip=True) if language else 'unknown'
    stars = repo.find('a', attrs={'href': re.compile('\/stargazers')})
    stars = int(stars.get_text(strip=True).replace(',', '')) if stars else 0
    print(name, language, stars)
```

Running this will output:

```
macuyiko.github.io HTML 0
blog JavaScript 1
minecraft-python JavaScript 14
[...]
```

As an extra exercise, try adapting this code to scrape out all pages in case the repositories page is paginated (as is the case for Google's account).

As a final add-on, you'll note that user pages like *https://github.com/Macuyiko?tab=repositories* also come with a short bio, including (in some cases) an e-mail address. However, this e-mail address is only visible once we log in to GitHub. In what follows, we'll try to get out this information as well.

Warning   This practice of hunting for a highly starred GitHub profile and extracting the contact information is frequently applied by recruitment firms. This being said, do note that we're now going to log in to GitHub and that we're crossing the boundary between public and private information. Consider this a practice exercise illustrating how you can do so in Python. Keeping the legal aspects in mind, you're advised to only scrape out your own profile information and to not set up this kind of scrapers on a large scale before knowing what you're getting into. Refer back to the chapter on legal concerns for the details regarding the legality of scraping.

You will need to create a GitHub profile in case you haven't done so already. Let us start by getting out the login form from the login page:

```python
import requests
from bs4 import BeautifulSoup

session = requests.Session()

url = 'https://github.com/{}'
username = 'Macuyiko'

# Visit the login page
r = session.get(url.format('login'))
html_soup = BeautifulSoup(r.text, 'html.parser')

form = html_soup.find(id='login')
print(form)
```

Running this will output:

```
<div class="auth-form px-3" id="login"> <!-- '"` -->
<!-- </textarea></xmp> --></div>
```

This is not exactly what we expected. If we take a look at the page source, we see that the page is formatted somewhat strangely:

```
<div class="auth-form px-3" id="login">

    <!-- '"` --><!-- </textarea></xmp> --></option></form>

    <form accept-charset="UTF-8" action="/session" method="post">
    <div style="margin:0;padding:0;display:inline">
    <input name="utf8" type="hidden" value="&#x2713;" />
    <input name="authenticity_token" type="hidden" value="AtuMda[...]zw==" />
    </div>

    <div class="auth-form-header p-0">
        <h1>Sign in to GitHub</h1>
    </div>

    <div id="js-flash-container">
</div>
[...]

</form>
```

The following modification makes sure we get out the forms in the page:

```python
import requests
from bs4 import BeautifulSoup

session = requests.Session()

url = 'https://github.com/{}'
username = 'Macuyiko'

# Visit the login page
r = session.get(url.format('login'))
html_soup = BeautifulSoup(r.text, 'html.parser')
```

```
data = {}
for form in html_soup.find_all('form'):
    # Get out the hidden form fields
    for inp in form.select('input[type=hidden]'):
        data[inp.get('name')] = inp.get('value')

# SET YOUR LOGIN DETAILS:
data.update({'login': '', 'password': ''})

print('Going to login with the following POST data:')
print(data)

if input('Do you want to login (y/n): ') == 'y':
    # Perform the login
    r = session.post(url.format('session'), data=data)
    # Get the profile page
    r = session.get(url.format(username))
    html_soup = BeautifulSoup(r.text, 'html.parser')
    user_info = html_soup.find(class_='vcard-details')
    print(user_info.text)
```

Even Browsers Have Bugs   If you've been using Chrome, you might wonder why you're not seeing the form data when following along with the login process using Chrome's Developer Tools. The reason is that Chrome contains a bug that will prevent form data from appearing in Developer Tools when the status code of the POST corresponds with a redirect. The POST data is still being sent; however, you just won't see it in the Developer Tools tab. This bug will probably be fixed by the time you're reading this, but it just goes to show that bugs appear in browsers as well.

Running this will output:

```
Going to login with the following POST data:
{'utf8': 'V',
 'authenticity_token': 'zgndmzes [...]',
 'login': 'YOUR_USER_NAME',
 'password': 'YOUR_PASSWORD'}
```

```
Do you want to login (y/n): y
```

KU Leuven

Belgium

macuyiko@gmail.com

http://blog.macuyiko.com

Plain Text Passwords   It goes without saying that hard-coding your password in plain text in Python files (and other programs, for that matter) is not advisable for real-life scripts. In a real deployment setting, where your code might get shared with others, make sure to modify your script so that it retrieves stored credentials from a secure data store (e.g., from the operating system environment variables, a file, or a database, preferably encrypted). Take a look at the "secureconfig" library available in pip, for example, on how to do so.

9.6 Scraping Mortgage Rates

We're going to scrape Barclays' mortgage simulator available at *https://www.barclays. co.uk/mortgages/mortgage-calculator/*. There isn't a particular reason why we pick this financial services provider, other than the fact that it applies some interesting techniques that serve as a nice illustration.

Take some time to explore the site a bit (using "What would it cost?"). We're asked to fill in a few parameters, after which we get an overview of possible products that we'd like to scrape out.

If you follow along with your browser's developer tools, you'll note that a POST request is being made to *https://www.barclays.co.uk/dss/service/co.uk/ mortgages/costcalculator/productservice*, with an interesting property: the JavaScript on the page performing the POST is using an "application/json" value for the "Content-Type" header and is including the POST data as plain JSON; see Figure 9-2. Depending on requests' `data` argument will not work in this case as it will encode the POST data. Instead, we need to use the `json` argument, which will basically instruct requests to format the POST data as JSON.

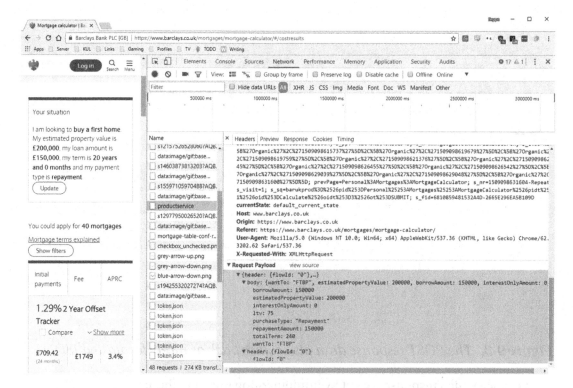

Figure 9-2. *The Barclays mortgage simulator submits a POST request using JavaScript and embeds the request data in a JSON format*

Additionally, you'll note that the result page is formatted as a relatively complex-looking table (with "Show more" links for every entry), though the response returned by the POST request looks like a nicely formatted JSON object; see Figure 9-3, so we might not even need Beautiful Soup here to access this "internal API".

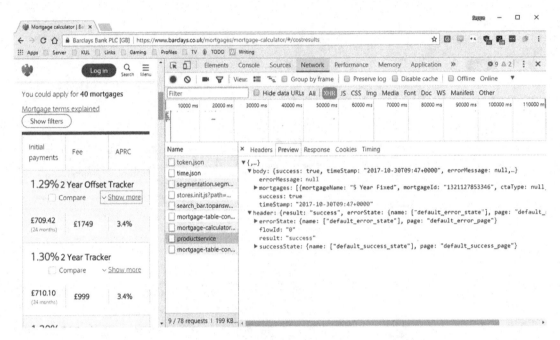

Figure 9-3. *The POST response data also comes back as nicely formatted JSON*

Let's see which response we get by implementing this in Python:

```python
import requests

url = 'https://www.barclays.co.uk/dss/service/co.uk/mortgages/' + \
      'costcalculator/productservice'

session = requests.Session()

estimatedPropertyValue = 200000
repaymentAmount = 150000
months = 240
data = {"header": {"flowId":"0"},
        "body":
        {"wantTo":"FTBP",
         "estimatedPropertyValue":estimatedPropertyValue,
         "borrowAmount":repaymentAmount,
         "interestOnlyAmount":0,
```

```
            "repaymentAmount":repaymentAmount,
            "ltv":round(repaymentAmount/estimatedPropertyValue*100),
            "totalTerm":months,
            "purchaseType":"Repayment"}}
```

```
r = session.post(url, json=data)
```

print(r.json())

Running this will output:

```
{'header':
{'result': 'error', 'systemError':
  {'errorCode': 'DSS_SEF001', 'type': 'E',
   'severity': 'FRAMEWORK',
   'errorMessage': 'State details not found in database',
   'validationErrors': [],
   'contentType': 'application/json', 'channel': '6'}
}}
```

That doesn't look too good. Remember that, when we don't get back the results we expect, there are various things we can do:

- Check whether we've forgotten to include some cookies. For example, we might need to visit the entry page first, or there might be cookies set by JavaScript. If you inspect the request in your browser, you'll note that there are a lot of cookies present.

- Check whether we've forgotten to include some headers, or whether we need to spoof some.

- If all else fails, resort to Selenium to implement a full browser.

In this particular situation, there are a lot of cookies being included in the request, some of which are set through normal "Set-Cookie" headers, though many are also set through a vast collection of JavaScript files included by the page. These would certainly be hard to figure out, as the JavaScript is obfuscated. There are, however, some interesting headers that are being set and included by JavaScript in the POST request, which do seem to be connected to the error message. Let's try including these, as well as spoofing the "User-Agent" and "Referer" headers:

```python
import requests

url = 'https://www.barclays.co.uk/dss/service/co.uk/mortgages/' + \
      'costcalculator/productservice'

session = requests.Session()

session.headers.update({
    # These are non-typical headers, let's include them
    'currentState': 'default_current_state',
    'action': 'default',
    'Origin': 'https://www.barclays.co.uk',
    # Spoof referer, user agent, and X-Requested-With
    'Referer': 'https://www.barclays.co.uk/mortgages/mortgage-calculator/',
    'User-Agent': 'Mozilla/5.0 (Windows NT 10.0; Win64; x64)
    AppleWebKit/537.36 ' + ' (KHTML, like Gecko) Chrome/62.0.3202.62
    Safari/537.36',
    'X-Requested-With': 'XMLHttpRequest',
    })
estimatedPropertyValue = 200000
repaymentAmount = 150000
months = 240
data = {"header": {"flowId":"0"},
        "body":
        {"wantTo":"FTBP",
         "estimatedPropertyValue":estimatedPropertyValue,
         "borrowAmount":repaymentAmount,
         "interestOnlyAmount":0,
         "repaymentAmount":repaymentAmount,
         "ltv":round(repaymentAmount/estimatedPropertyValue*100),
         "totalTerm":months,
         "purchaseType":"Repayment"}}

r = session.post(url, json=data)
```

```
# Only print the header to avoid text overload
print(r.json()['header'])
```

This seems to work! In this case, it in fact turns out we didn't have to include any cookies at all. We can now clean up this code:

```python
import requests

def get_mortgages(estimatedPropertyValue, repaymentAmount, months):
    url = 'https://www.barclays.co.uk/dss/service/' + \
            'co.uk/mortgages/costcalculator/productservice'
    headers = {
        # These are non-typical headers, let's include them
        'currentState': 'default_current_state',
        'action': 'default',
        'Origin': 'https://www.barclays.co.uk',
        # Spoof referer, user agent, and X-Requested-With
        'Referer': 'https://www.barclays.co.uk/mortgages/mortgage-
        calculator/',
        'User-Agent': 'Mozilla/5.0 (Windows NT 10.0; Win64; x64)
        AppleWebKit/537.36 ' + ' (KHTML, like Gecko) Chrome/62.0.3202.62
        Safari/537.36',
        'X-Requested-With': 'XMLHttpRequest',
        }
    data = {"header": {"flowId":"0"},
            "body":
            {"wantTo":"FTBP",
             "estimatedPropertyValue":estimatedPropertyValue,
             "borrowAmount":repaymentAmount,
             "interestOnlyAmount":0,
             "repaymentAmount":repaymentAmount,
             "ltv":round(repaymentAmount/estimatedPropertyValue*100),
             "totalTerm":months,
             "purchaseType":"Repayment"}}
```

```
    r = requests.post(url, json=data, headers=headers)
    results = r.json()
    return results['body']['mortgages']
mortgages = get_mortgages(200000, 150000, 240)

# Print the first mortgage info
print(mortgages[0])
```

Running this will output:

```
{'mortgageName': '5 Year Fixed', 'mortgageId': '1321127853346',
 'ctaType': None, 'uniqueId': '590b357e295b0377d0fb607b',
 'mortgageType': 'FIXED',
 'howMuchCanBeBorrowedNote': '95% (max) of the value of your home',
 'initialRate': 4.99, 'initialRateTitle': '4.99%',
 'initialRateNote': 'until 31st January 2023',
 [...]
```

9.7 Scraping and Visualizing IMDB Ratings

The next series of examples moves on toward including some more data science-oriented use cases. We're going to start simple by scraping a list of reviews for episodes of a TV series, using IMDB (the Internet Movie Database). We'll use *Game of Thrones* as an example, the episode list for which can be found at *http://www.imdb.com/title/ tt0944947/episodes*. Note that IMDB's overview is spread out across multiple pages (per season or per year), so we iterate over the seasons we want to retrieve using an extra loop:

```
import requests
from bs4 import BeautifulSoup

url = 'http://www.imdb.com/title/tt0944947/episodes'

episodes = []
ratings = []
```

```
# Go over seasons 1 to 7
for season in range(1, 8):
    r = requests.get(url, params={'season': season})
    soup = BeautifulSoup(r.text, 'html.parser')
    listing = soup.find('div', class_='eplist')
    for epnr, div in enumerate(listing.find_all('div', recursive=False)):
        episode = "{}.{}".format(season, epnr + 1)
        rating_el = div.find(class_='ipl-rating-star__rating')
        rating = float(rating_el.get_text(strip=True))
        print('Episode:', episode, '-- rating:', rating)
        episodes.append(episode)
        ratings.append(rating)
```

We can then plot the scraped ratings using "matplotlib," a well-known plotting library for Python that can be easily installed using pip:

```
pip install -U matplotlib
```

Plotting with Python   Of course, you could also reproduce the plot below using, for example, Excel, but this example serves as a gentle introduction as we'll continue to use matplotlib for some later examples as well. Note that this is certainly not the only—or even most user-friendly—plotting library for Python, though it remains one of the most prevalent ones. Take a look at Seaborn (*https://seaborn.pydata.org/*), Altair (*https://altair-viz.github.io/*) and ggplot (*http://ggplot.yhathq.com/*) for some other excellent libraries.

Adding in the following lines to our script plots the results in a simple bar chart, as shown in Figure 9-4.

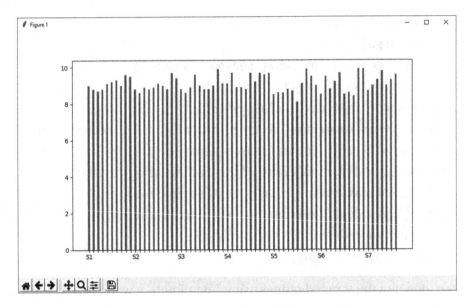

Figure 9-4. Plotting IMDB ratings per episode using "matplotlib"

```python
import matplotlib.pyplot as plt

episodes = ['S' + e.split('.')[0] if int(e.split('.')[1]) == 1 else '' \
                        for e in episodes]

plt.figure()
positions = [a*2 for a in range(len(ratings))]
plt.bar(positions, ratings, align='center')
plt.xticks(positions, episodes)
plt.show()
```

9.8 Scraping IATA Airline Information

We're going to scrape airline information using the search form available at *http://www.iata.org/publications/Pages/code-search.aspx*. This is an interesting case to illustrate the "nastiness" of some websites, even though the form we want to use looks incredibly simple (there's only one drop-down and one text field visible on the page). As the URL already shows, the web server driving this page is built on ASP.NET (".aspx"), which has very peculiar opinions about how it handles form data.

It is a good idea to try submitting this form using your browser and taking a look at what happens using its developer tools. As you can see in Figure 9-5, it seems that a lot of form data get included in the POST request — much more than our two fields.

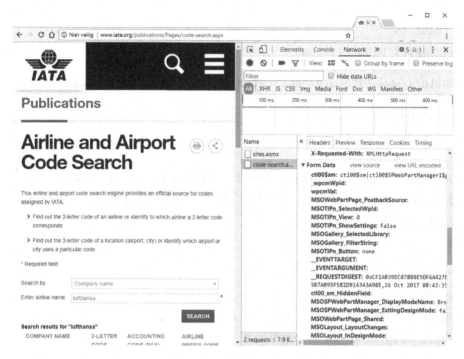

Figure 9-5. *Submitting the IATA form includes lots of form data.*

Certainly, it does not look feasible to manually include all these fields in our Python script. The "__VIEWSTATE" field, for instance, holds session information that changes for every request. Even some names of fields seem to include parts of which we can't really be sure that they wouldn't change in the future, causing our script to break. In addition, it seems that we also need to keep track of cookies as well. Finally, take a look at the response content that comes back from the POST request. This looks like a partial response (which will be parsed and shown by JavaScript) instead of a full HTML page:

```
1|#||4|1330|updatePanel|ctl00_SPWebPartManager1_g_e3b09024_878e
[...]
MSOSPWebPartManager_StartWebPartEditingName|false|5|hiddenField|
MSOSPWebPartManager_EndWebPartEditing|false|
```

To handle these issues, we're going to try to make our code as robust as possible. First, we'll start by performing a GET request to the search page, using requests' sessions mechanism. Next, we'll use Beautiful Soup to get out all the form elements with their names and values:

```python
import requests
from bs4 import BeautifulSoup

url = 'http://www.iata.org/publications/Pages/code-search.aspx'

session = requests.Session()
# Spoof the user agent as a precaution
session.headers.update({
    'User-Agent' : 'Mozilla/5.0 (Windows NT 10.0; Win64; x64)
    AppleWebKit/537.36 ' + ' (KHTML, like Gecko) Chrome/62.0.3202.62
    Safari/537.36'
    })

# Get the search page
r = session.get(url)
html_soup = BeautifulSoup(r.text, 'html.parser')
form = html_soup.find(id='aspnetForm')

# Get the form fields
data = {}
for inp in form.find_all(['input', 'select']):
    name = inp.get('name')
    value = inp.get('value')
    if not name:
        continue
    data[name] = value if value else ''

print(data, end='\n\n\n')
```

This will output the following:

```
{'_wpcmWpid': '',
 'wpcmVal': '',
 'MSOWebPartPage_PostbackSource': '',
 'MSOTlPn_SelectedWpId': '',
```

```
'MSOTlPn_View': '0',
'MSOTlPn_ShowSettings': 'False',
'MSOGallery_SelectedLibrary': '',
'MSOGallery_FilterString': '',
'MSOTlPn_Button': 'none',
'__EVENTTARGET': '',
'__EVENTARGUMENT': '',
[...]
```

Next, we'll use the collected form data to perform a POST request. We do have to make sure to set the correct values for the drop-down and text box, however. We add the following lines to our script:

```
# Set our desired search query
for name in data.keys():
    # Search by
    if 'ddlImLookingFor' in name:
        data[name] = 'ByAirlineName'
    # Airline name
    if 'txtSearchCriteria' in name:
        data[name] = 'Lufthansa'

# Perform a POST
r = session.post(url, data=data)
print(r.text)
```

Strangely enough, contrary to what's happening in the browser, the POST request does return a full HTML page here, instead of a partial result. This is not too bad, as we can now use Beautiful Soup to fetch the table of results.

Instead of parsing this table manually, we'll use a popular data science library for tabular data wrangling called "pandas," which comes with a helpful "HTML table to data frame" method built in. The library is easy to install using pip:

```
pip install -U pandas
```

To parse out HTML, pandas relies on "lxml" by default and falls back to Beautiful Soup with "html5lib" in case "lxml" cannot be found. To make sure "lxml" is available, install it with:

```
pip install -U lxml
```

The full script can now be organized to look as follows:

```python
import requests
from bs4 import BeautifulSoup
import pandas

url = 'http://www.iata.org/publications/Pages/code-search.aspx'

def get_results(airline_name):
    session = requests.Session()
    # Spoof the user agent as a precaution
    session.headers.update({
        'User-Agent' : 'Mozilla/5.0 (Windows NT 10.0; Win64; x64)
        AppleWebKit/537.36 ' + ' (KHTML, like Gecko) Chrome/62.0.3202.62
        Safari/537.36'
        })
    r = session.get(url)
    html_soup = BeautifulSoup(r.text, 'html.parser')
    form = html_soup.find(id='aspnetForm')
    data = {}
    for inp in form.find_all(['input', 'select']):
        name = inp.get('name')
        value = inp.get('value')
        if not name:
            continue
        if 'ddlImLookingFor' in name:
            value = 'ByAirlineName'
        if 'txtSearchCriteria' in name:
            value = airline_name
        data[name] = value if value else ''

    r = session.post(url, data=data)
    html_soup = BeautifulSoup(r.text, 'html.parser')
    table = html_soup.find('table', class_='datatable')
    df = pandas.read_html(str(table))
    return df

df = get_results('Lufthansa')
print(df)
```

Running this will output:

```
[                                0    1      2      3
0            Deutsche Lufthansa AG  LH  220.0  220.0
1                Lufthansa Cargo AG  LH    NaN   20.0
2           Lufthansa CityLine GmbH  CL  683.0  683.0
3 Lufthansa Systems GmbH & Co. KG  S1    NaN    NaN]
```

The equivalent Selenium code looks as follows:

```python
import pandas
from selenium import webdriver
from selenium.webdriver.support.ui import Select

url = 'http://www.iata.org/publications/Pages/code-search.aspx'

driver = webdriver.Chrome()
driver.implicitly_wait(10)

def get_results(airline_name):
    driver.get(url)
    # Make sure to select the right part of the form
    # This will make finding the elements easier
    # as #aspnetForm wraps the whole page, including
    # the search box
    form_div = driver.find_element_by_css_selector('#aspnetForm
    .iataStandardForm')
    select = Select(form_div.find_element_by_css_selector('select'))
    select.select_by_value('ByAirlineName')
    text = form_div.find_element_by_css_selector('input[type=text]')
    text.send_keys(airline_name)
    submit = form_div.find_element_by_css_selector('input[type=submit]')
    submit.click()
    table = driver.find_element_by_css_selector('table.datatable')
    table_html = table.get_attribute('outerHTML')
    df = pandas.read_html(str(table_html))
    return df
```

```
df = get_results('Lufthansa')
print(df)

driver.quit()
```

There's still one mystery we have to solve: remember that the POST request as made by requests returns a full HTML page, instead of a partial result as we observed in the browser. How does the server figure out how to differentiate between both types of results? The answer lies in the way the search form is submitted. In requests, we perform a simple POST request with a minimal amount of headers. On the live page, however, the form submission is handled by JavaScript, which will perform the actual POST request and will parse out the partial results to show them. To indicate to the server that it is JavaScript making the request, two headers are included in the request, which we can spoof in requests as well. If we modify our code as follows, you will indeed also obtain the same partial result:

```
# Include headers to indicate that we want a partial result
session.headers.update({
    'X-MicrosoftAjax'  : 'Delta=true',
    'X-Requested-With' : 'XMLHttpRequest',
    'User-Agent' : 'Mozilla/5.0 (Windows NT 10.0; Win64; x64)
    AppleWebKit/537.36 ' + ' (KHTML, like Gecko) Chrome/62.0.3202.62
    Safari/537.36'
})
```

9.9 Scraping and Analyzing Web Forum Interactions

In this example, we're going to scrape web forum posts available at *http://bpbasecamp.freeforums.net/board/27/gear-closet* (a forum for backpackers and hikers) to get an idea about who the most active users are and who is frequently interacting with whom. We're going to keep a tally of interactions that will be constructed as follows:

- The first post in a "thread" is not "replying" to anyone, so we won't consider this as an interaction,

- The next posts in a thread can optionally include one or more quote blocks, which indicate that the poster is directly replying to another user, which we'll regard as such,

- If a post does not include any quote blocks, we'll just assume the post to be a reply to the original poster. This might not necessarily be the case, and users will oftentimes use little pieces of text such as "^^" to indicate they're replying to the direct previous poster, but we're going to keep it simple in this example (feel free to modify the scripts accordingly to your definition of "interaction," however).

Let's get started. First, we're going to extract a list of threads given a forum URL:

```python
import requests
import re
from bs4 import BeautifulSoup

def get_forum_threads(url, max_pages=None):
    page = 1
    threads = []
    while not max_pages or page <= max_pages:
        print('Scraping forum page:', page)
        r = requests.get(url, params={'page': page})
        soup = BeautifulSoup(r.text, 'html.parser')
        content = soup.find(class_='content')
        links = content.find_all('a', attrs={'href': re.compile
('^\/thread\/')})
        threads_on_page = [a.get('href') for a in links \
                if a.get('href') and not 'page=' in a.get('href')]
        threads += threads_on_page
        page += 1
        next_page = soup.find('li', class_='next')
        if 'state-disabled' in next_page.get('class'):
            break
    return threads

url = 'http://bpbasecamp.freeforums.net/board/27/gear-closet'

threads = get_forum_threads(url, max_pages=5)
print(threads)
```

Note that we have to be a bit clever here regarding pagination. This forum will continue to return the last page, even when supplying higher than maximum page numbers as the URL parameter, so that we can check whether an item with the class "next" also has the class "state-disabled" to determine whether we've reached the end of the thread list. Since we only want thread links corresponding with the first page, we remove all links that have "page=" in their URL as well. In the example, we also decide to limit ourselves to five pages only. Running this will output:

```
Scraping forum page: 1
Scraping forum page: 2
Scraping forum page: 3
Scraping forum page: 4
Scraping forum page: 5
['/thread/2131/before-asking-which-pack-boot', [...] ]
```

For every thread, we now want to get out a list of posts. We can try this out with one thread first:

```python
import requests
import re
from urllib.parse import urljoin
from bs4 import BeautifulSoup

def get_thread_posts(url, max_pages=None):
    page = 1
    posts = []
    while not max_pages or page <= max_pages:
        print('Scraping thread url/page:', url, page)
        r = requests.get(url, params={'page': page})
        soup = BeautifulSoup(r.text, 'html.parser')
        content = soup.find(class_='content')
        for post in content.find_all('tr', class_='item'):
            user = post.find('a', class_='user-link')
            if not user:
                # User might be deleted, skip...
                continue
```

```
            user = user.get_text(strip=True)
            quotes = []
            for quote in post.find_all(class_='quote_header'):
                quoted_user = quote.find('a', class_='user-link')
                if quoted_user:
                    quotes.append(quoted_user.get_text(strip=True))
            posts.append((user, quotes))
        page += 1
        next_page = soup.find('li', class_='next')
        if 'state-disabled' in next_page.get('class'):
            break
    return posts
url = 'http://bpbasecamp.freeforums.net/board/27/gear-closet'
thread = '/thread/2131/before-asking-which-pack-boot'

thread_url = urljoin(url, thread)
posts = get_thread_posts(thread_url)
print(posts)
```

Running this will output a list with every element being a tuple containing the poster's name and a list of users that are quoted in the post:

```
Scraping thread url/page:                                          ↵
    http://bpbasecamp.freeforums.net/thread/2131/before-asking-which-pack-boot 1
Scraping thread url/page:                                          ↵
    http://bpbasecamp.freeforums.net/thread/2131/before-asking-which-pack-boot 2
[('almostthere', []), ('trinity', []), ('paula53', []),            ↵
    ('toejam', ['almostthere']), ('stickman', []), ('tamtrails', []),   ↵
    ('almostthere', ['tamtrails']), ('kayman', []), ('almostthere',     ↵
    ['kayman']), ('lanceman', []), ('trinity', ['trinity']),            ↵
     ('Christian', ['almostthere']), ('pollock', []), ('mitsmit', []),  ↵
    ('intothewild', []), ('Christian', []), ('softskull', []), ('argus',↵
    []),('lyssa7', []), ('kevin', []), ('greenwoodsuncharted', [])]
```

231

By putting both of these functions together, we get the script below. We'll use Python's "pickle" module to store our scraped results so that we don't have to rescrape the forum over and over again:

```python
import requests
import re
from urllib.parse import urljoin
from bs4 import BeautifulSoup
import pickle

def get_forum_threads(url, max_pages=None):
    page = 1
    threads = []
    while not max_pages or page <= max_pages:
        print('Scraping forum page:', page)
        r = requests.get(url, params={'page=': page})
        soup = BeautifulSoup(r.text, 'html.parser')
        content = soup.find(class_='content')
        links = content.find_all('a', attrs={'href': re.compile
        ('^\/thread\/')})
        threads_on_page = [a.get('href') for a in links \
                if a.get('href') and not 'page' in a.get('href')]
        threads += threads_on_page
        page += 1
        next_page = soup.find('li', class_='next')
        if 'state-disabled' in next_page.get('class'):
            break
    return threads
def get_thread_posts(url, max_pages=None):
    page = 1
    posts = []
```

```python
    while not max_pages or page <= max_pages:
        print('Scraping thread url/page:', url, page)
        r = requests.get(url, params={'page': page})
        soup = BeautifulSoup(r.text, 'html.parser')
        content = soup.find(class_='content')
        for post in content.find_all('tr', class_='item'):
            user = post.find('a', class_='user-link')
            if not user:
                # User might be deleted, skip...
                continue
            user = user.get_text(strip=True)
            quotes = []
            for quote in post.find_all(class_='quote_header'):
                quoted_user = quote.find('a', class_='user-link')
                if quoted_user:
                    quotes.append(quoted_user.get_text(strip=True))
            posts.append((user, quotes))
        page += 1
        next_page = soup.find('li', class_='next')
        if 'state-disabled' in next_page.get('class'):
            break
    return posts

url = 'http://bpbasecamp.freeforums.net/board/27/gear-closet'

threads = get_forum_threads(url, max_pages=5)
all_posts = []

for thread in threads:
    thread_url = urljoin(url, thread)
    posts = get_thread_posts(thread_url)
    all_posts.append(posts)

with open('forum_posts.pkl', "wb") as output_file:
    pickle.dump(all_posts, output_file)
```

Next, we can load the results and visualize them in a heat map. We're going to use "pandas," "numpy," and "matplotlib" to do so, all of which can be installed through pip (if you've already installed pandas and matplotlib by following the previous examples, there's nothing else you have to install):

```
pip install -U pandas
pip install -U numpy
pip install -U matplotlib
```

Let's start by visualizing the first thread only (shown in the output fragment of the scraper above):

```python
import pickle
import numpy as np
import pandas as pd
import matplotlib.pyplot as plt

# Load our stored results
with open('forum_posts.pkl', "rb") as input_file:
    posts = pickle.load(input_file)

def add_interaction(users, fu, tu):
    if fu not in users:
        users[fu] = {}
    if tu not in users[fu]:
        users[fu][tu] = 0
    users[fu][tu] += 1

# Create interactions dictionary
users = {}
for thread in posts:
    first_one = None
    for post in thread:
        user = post[0]
        quoted = post[1]
        if not first_one:
            first_one = user
        elif not quoted:
```

```
            add_interaction(users, user, first_one)
        else:
            for qu in quoted:
                add_interaction(users, user, qu)
    # Stop after the first thread
    break

df = pd.DataFrame.from_dict(users, orient='index').fillna(0)

heatmap = plt.pcolor(df, cmap='Blues')
y_vals = np.arange(0.5, len(df.index), 1)
x_vals = np.arange(0.5, len(df.columns), 1)
plt.yticks(y_vals, df.index)
plt.xticks(x_vals, df.columns, rotation='vertical')
for y in range(len(df.index)):
    for x in range(len(df.columns)):
        if df.iloc[y, x] == 0:
            continue
        plt.text(x + 0.5, y + 0.5, '%.0f' % df.iloc[y, x],
                 horizontalalignment='center',
                 verticalalignment='center')
plt.show()
```

This will provide you with a result as shown in Figure 9-6. As you can see, various users are replying to the original poster, and the original poster is also quoting some other users.

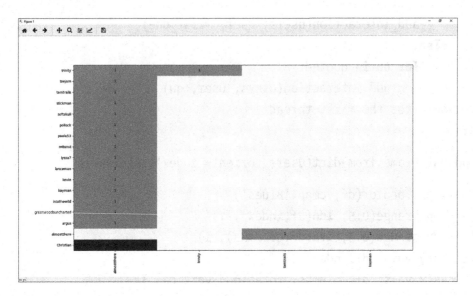

Figure 9-6. *Visualizing user interactions for one forum thread*

There are various ways to play around with this visualization. Figure 9-7, for instance, shows the user interactions over all forum threads, but only taking into account direct quotes.

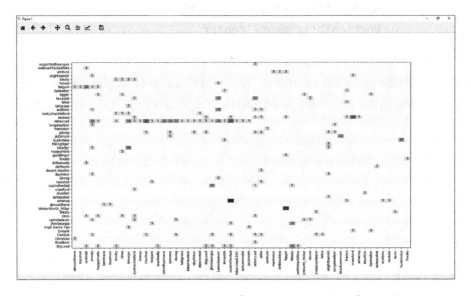

Figure 9-7. *Visualizing user interactions (direct quotes only) for all scraped forum threads*

9.10 Collecting and Clustering a Fashion Data Set

In this example, we're going to use Zalando (a popular Swedish web shop) to fetch a collection of images of fashion products and cluster them using t-SNE.

Check the API   Note that Zalando also exposes an easy to use API (see *https://github.com/zalando/shop-api-documentation/wiki/Api-introduction* for the documentation). At the time of writing, the API does not require authentication, though this is scheduled to change in the near future, requiring users to register to get an API access token. Since we'll only fetch images here, we'll not bother to register, though in a proper "app," using the API option would certainly be recommended.

Our first script downloads images and stores them in a directory; see Figure 9-8:

```python
import requests
import os, os.path
from bs4 import BeautifulSoup
from urllib.parse import urljoin, urlparse

store = 'images'
if not os.path.exists(store):
    os.makedirs(store)

url = 'https://www.zalando.co.uk/womens-clothing-dresses/'
pages_to_crawl = 15

def download(url):
    r = requests.get(url, stream=True)
    filename = urlparse(url).path.split('/')[-1]
    print('Downloading to:', filename)
    with open(os.path.join(store, filename), 'wb') as the_image:
        for byte_chunk in r.iter_content(chunk_size=4096*4):
            the_image.write(byte_chunk)
```

```
for p in range(1, pages_to_crawl+1):
    print('Scraping page:', p)
    r = requests.get(url, params={'p' : p})
    html_soup = BeautifulSoup(r.text, 'html.parser')
    for img in html_soup.select('#z-nvg-cognac-root z-grid-item img'):
        img_src = img.get('src')
        if not img_src:
            continue
        img_url = urljoin(url, img_src)
        download(img_url)
```

Figure 9-8. *A collection of scraped dress images*

Next, we'll use the t-SNE clustering algorithm to cluster the photos. t-SNE is a relatively recent dimensionality reduction technique that is particularly well-suited for the visualization of high-dimensional data sets, like images. You can read about the technique at *https://lvdmaaten.github.io/tsne/*. We're going to use "scikit-learn" together with "matplotlib," "scipy," and "numpy," all of which are libraries that are familiar to data scientists and can be installed through pip:

```
pip install -U matplotlib
pip install -U scikit-learn
pip install -U numpy
pip install -U scipy
```

Our clustering script looks as follows:

```python
import os.path
import numpy as np
import matplotlib.pyplot as plt
from matplotlib import offsetbox
from sklearn import manifold
from scipy.misc import imread
from glob import iglob

store = 'images'

image_data = []
for filename in iglob(os.path.join(store, '*.jpg')):
    image_data.append(imread(filename))

image_np_orig = np.array(image_data)
image_np = image_np_orig.reshape(image_np_orig.shape[0], -1)

def plot_embedding(X, image_np_orig):
    # Rescale
    x_min, x_max = np.min(X, 0), np.max(X, 0)
    X = (X - x_min) / (x_max - x_min)
    # Plot images according to t-SNE position
    plt.figure()
    ax = plt.subplot(111)
    for i in range(image_np.shape[0]):
        imagebox = offsetbox.AnnotationBbox(
            offsetbox=offsetbox.OffsetImage(image_np_orig[i], zoom=.1),
            xy=X[i],
            frameon=False)
        ax.add_artist(imagebox)
```

```
print("Computing t-SNE embedding")

tsne = manifold.TSNE(n_components=2, init='pca')
X_tsne = tsne.fit_transform(image_np)

plot_embedding(X_tsne, image_np_orig)
plt.show()
```

This code works as follows. First, we load all the images (using `imread`) and convert them to a numpy array. The `reshape` function makes sure that we get a n x $3m$ matrix, with n the number of images and m the number of pixels per image, instead of an n x r x g x b tensor, with r, g, and b the pixel values for the red, green, and blue channels respectively. After constructing the t-SNE embedding, we plot the images with their calculated x and y coordinates using matplotlib, resulting in an image like shown in Figure 9-9 (using about a thousand scraped photos). As can be seen, the clustering here is primarily driven by the saturation and intensity of the colors in the image.

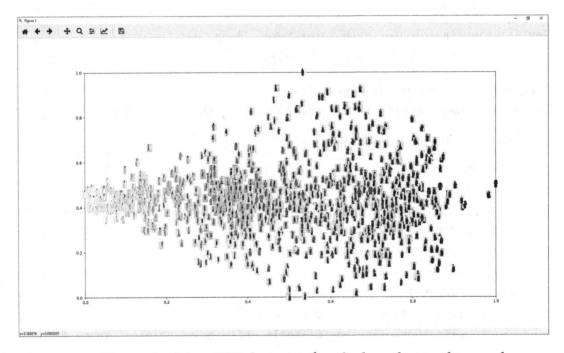

Figure 9-9. *The result of the t-SNE clustering (applied on about a thousand photos)*

Image Sizes We're lucky that all of the images we've scraped have the same width and height. If this would not be the case, we'd first have to apply a resizing to make sure every image will lead to a vector in the data set with equal length.

9.11 Sentiment Analysis of Scraped Amazon Reviews

We're going to scrape a list of Amazon reviews with their ratings for a particular product. We'll use a book with plenty of reviews, say *Learning Python by Mark Lutz*, which can be found at *https://www.amazon.com/Learning-Python-5th-Mark-Lutz/dp/1449355730/*. If you click through "See all customer reviews," you'll end up at *https://www.amazon.com/Learning-Python-5th-Mark-Lutz/product-reviews/1449355730/*. Note that this product has an id of "1449355730," and even using the URL *https://www.amazon.com/product-reviews/1449355730/*, without the product name, will work.

Simple URLs Playing around with URLs as we do here is always a good idea before writing your web scraper. Based on the above, we know that a given product identifier is enough to fetch the reviews page, without a need to figure out the exact URL, including the product name. Why then, does Amazon allow for both and does it default to including the product name? The reason is most likely search engine optimization (SEO). Search engines like Google prefer URLs with human-readable components included.

If you explore the reviews page, you'll note that the reviews are paginated. By browsing to other pages and following along in your browser's developer tools, we see that POST requests are being made (by JavaScript) to URLs looking like *https://www.amazon.com/ss/customer-reviews/ajax/reviews/get/ref=cm_cr_arp_d_paging_btm_2*, with the product id included in the form data, as well as some other form fields that look relatively easy to spoof. Let's see what we get in requests:

```
import requests
from bs4 import BeautifulSoup

review_url = 'https://www.amazon.com/ss/customer-reviews/ajax/reviews/get/'
product_id = '1449355730'
```

```python
session = requests.Session()
session.headers.update({
    'User-Agent' : 'Mozilla/5.0 (Windows NT 10.0; Win64; x64)
    AppleWebKit/537.36 ' + ' (KHTML, like Gecko) Chrome/62.0.3202.62
    Safari/537.36'
    })

session.get('https://www.amazon.com/product-reviews/{}/'.format(product_id))

def get_reviews(product_id, page):
    data = {
        'sortBy':'',
        'reviewerType':'all_reviews',
        'formatType':'',
        'mediaType':'',
        'filterByStar':'all_stars',
        'pageNumber':page,
        'filterByKeyword':'',
        'shouldAppend':'undefined',
        'deviceType':'desktop',
        'reftag':'cm_cr_getr_d_paging_btm_{}'.format(page),
        'pageSize':10,
        'asin':product_id,
        'scope':'reviewsAjax1'
        }
    r = session.post(review_url + 'ref=' + data['reftag'], data=data)
    return r.text

print(get_reviews(product_id, 1))
```

Note that we spoof the "User-Agent" header here. If we don't, Amazon will reply with a message requesting us to verify whether we're a human (you can copy the value for this header from your browser's developer tools). In addition, note the "scope" form field that we set to "reviewsAjax1." If you explore the reviews page in the browser, you'll see that the value of this field is in fact increased for each request, that is, "reviewsAjax1," "reviewsAjax2," and so on. We could decide to replicate this behavior as well — which we'd have to do in case Amazon would pick up on our tactics, though it does not seem to be necessary for the results to come back correctly.

Finally, note that the POST request does not return a full HTML page, but some kind of hand-encoded result that will be parsed (normally) by JavaScript:

```
["script",
 "if(window.ue) { ues('id','reviewsAjax1','FE738GN7GRDZK6Q09S9G');
 ues('t0','reviewsAjax1',new Date());
 ues('ctb','reviewsAjax1','1');
 uet('bb','reviewsAjax1'); }"
]
&&&
["update","#cm_cr-review_list",""]
&&&
["loaded"]
&&&
["append","#cm_cr-review_list","<div id=\"R3JQXR4EMWJ7AD\" data-
    hook=\"review\"class=\"a-section review\"><div id=\
    "customer_review-R3JQXR4EMWJ7AD\"class=\"a-section celwidget\">
    <div class=\"a-row\"><a class=\"a-link-normal\"title=\"5.0 out
    of 5 stars\"
[...]
```

Luckily, after exploring the reply a bit (feel free to copy-paste the full reply in a text editor and read through it), the structure seems easy enough to figure out:

- The reply is composed of several "instructions," formatted as a JSON list;
- The instructions themselves are separated by three ampersands, "&&&";
- The instructions containing the reviews start with an "append" string;
- The actual contents of the review are formatted as an HTML element and found on the third position of the list.

Let's adjust our code to parse the reviews in a structured format. We'll loop through all the instructions; convert them using the "json" module; check for "append" entries; and then use Beautiful Soup to parse the HTML fragment and get the review id, rating, title, and text. We'll also need a small regular expression to get out the rating, which is set as a class with a value like "a-start-1" to "a-star-5". We could use these as is, but simply getting "1" to "5" might be easier to work with later on, so we already perform a bit of cleaning here:

```python
import requests
import json
import re
from bs4 import BeautifulSoup

review_url = 'https://www.amazon.com/ss/customer-reviews/ajax/reviews/get/'
product_id = '1449355730'

session = requests.Session()
session.headers.update({
    'User-Agent' : 'Mozilla/5.0 (Windows NT 10.0; Win64; x64)
    AppleWebKit/537.36 ' + ' (KHTML, like Gecko) Chrome/62.0.3202.62
    Safari/537.36'
    })

session.get('https://www.amazon.com/product-reviews/{}/'.format(product_id))

def parse_reviews(reply):
    reviews = []
    for fragment in reply.split('&&&'):
        if not fragment.strip():
            continue
        json_fragment = json.loads(fragment)
        if json_fragment[0] != 'append':
            continue
        html_soup = BeautifulSoup(json_fragment[2], 'html.parser')
        div = html_soup.find('div', class_='review')
        if not div:
            continue
        review_id = div.get('id')
        title = html_soup.find(class_='review-title').get_text(strip=True)
        review = html_soup.find(class_='review-text').get_text(strip=True)
```

```
    # Find and clean the rating:
    review_cls = ' '.join(html_soup.find(class_='review-rating').
    get('class'))
    rating = re.search('a-star-(\d+)', review_cls).group(1)
    reviews.append({'review_id': review_id,
                    'rating': rating,
                    'title': title,
                    'review': review})
    return reviews

def get_reviews(product_id, page):
    data = {
        'sortBy':'',
        'reviewerType':'all_reviews',
        'formatType':'',
        'mediaType':'',
        'filterByStar':'all_stars',
        'pageNumber':page,
        'filterByKeyword':'',
        'shouldAppend':'undefined',
        'deviceType':'desktop',
        'reftag':'cm_cr_getr_d_paging_btm_{}'.format(page),
        'pageSize':10,
        'asin':product_id,
        'scope':'reviewsAjax1'
        }
    r = session.post(review_url + 'ref=' + data['reftag'], data=data)
    reviews = parse_reviews(r.text)
    return reviews

print(get_reviews(product_id, 1))
```

This works! The only thing left to do is to loop through all the pages, and store the reviews in a database using the "dataset" library. Luckily, figuring out when to stop looping is easy: once we do not get any reviews for a particular page, we can stop:

245

```
import requests
import json
import re
from bs4 import BeautifulSoup
import dataset

db = dataset.connect('sqlite:///reviews.db')

review_url = 'https://www.amazon.com/ss/customer-reviews/ajax/reviews/get/'
product_id = '1449355730'

session = requests.Session()
session.headers.update({
    'User-Agent' : 'Mozilla/5.0 (Windows NT 10.0; Win64; x64)
    AppleWebKit/537.36 ' + ' (KHTML, like Gecko) Chrome/62.0.3202.62
    Safari/537.36'
    })

session.get('https://www.amazon.com/product-reviews/{}/'.format(product_id))

def parse_reviews(reply):
    reviews = []
    for fragment in reply.split('&&&'):
        if not fragment.strip():
            continue
        json_fragment = json.loads(fragment)
        if json_fragment[0] != 'append':
            continue
        html_soup = BeautifulSoup(json_fragment[2], 'html.parser')
        div = html_soup.find('div', class_='review')
        if not div:
            continue
        review_id = div.get('id')
        review_cls = ' '.join(html_soup.find(class_='review-rating').
        get('class'))
        rating = re.search('a-star-(\d+)', review_cls).group(1)
        title = html_soup.find(class_='review-title').get_text(strip=True)
        review = html_soup.find(class_='review-text').get_text(strip=True)
```

```
        reviews.append({'review_id': review_id,
                        'rating': rating,
                        'title': title,
                        'review': review})
    return reviews

def get_reviews(product_id, page):
    data = {
        'sortBy':'',
        'reviewerType':'all_reviews',
        'formatType':'',
        'mediaType':'',
        'filterByStar':'all_stars',
        'pageNumber':page,
        'filterByKeyword':'',
        'shouldAppend':'undefined',
        'deviceType':'desktop',
        'reftag':'cm_cr_getr_d_paging_btm_{}'.format(page),
        'pageSize':10,
        'asin':product_id,
        'scope':'reviewsAjax1'
        }
    r = session.post(review_url + 'ref=' + data['reftag'], data=data)
    reviews = parse_reviews(r.text)
    return reviews

page = 1
while True:
    print('Scraping page', page)
    reviews = get_reviews(product_id, page)
    if not reviews:
        break
    for review in reviews:
        print(' -', review['rating'], review['title'])
        db['reviews'].upsert(review, ['review_id'])
    page += 1
```

This will output the following:

```
Scraping page 1
  - 5 let me try to explain why this 1600 page book may actually end    ↵
      up saving you a lot of time and making you a better Python progra
  - 5 Great start, and written for the novice
  - 5 Best teacher of software development
  - 5 Very thorough
  - 5 If you like big thick books that deal with a lot of ...
  - 5 Great book, even for the experienced python programmer
  - 5 Good Tutorial; you'll learn a lot.
  - 2 Takes too many pages to explain even the most simpliest ...
  - 3 If I had a quarter for each time he says something like "here's   ↵
      an intro to X
  - 4 it almost seems better suited for a college class
[...]
```

Now that we have a database containing the reviews, let's do something fun with these. We'll run a sentiment analysis algorithm over the reviews (providing a sentiment score per review), which we can then plot over the different ratings given to inspect the correlation between a rating and the sentiment in the text. To do so, we'll use the "vaderSentiment" library, which can simply be installed using pip. We'll also need to install the "nltk" (Natural Language Toolkit) library:

```
pip install -U vaderSentiment
pip install -U nltk
```

Using the vaderSentiment library is pretty simple for a single sentence:

```python
from vaderSentiment.vaderSentiment import SentimentIntensityAnalyzer

analyzer = SentimentIntensityAnalyzer()

sentence = "I'm really happy with my purchase"
vs = analyzer.polarity_scores(sentence)

print(vs)
# Shows: {'neg': 0.0, 'neu': 0.556, 'pos': 0.444, 'compound': 0.6115}
```

To get the sentiment for a longer piece of text, a simple approach is to calculate the sentiment score per sentence and average this over all the sentences in the text, like so:

```
from vaderSentiment.vaderSentiment import SentimentIntensityAnalyzer
from nltk import tokenize

analyzer = SentimentIntensityAnalyzer()

paragraph = """
    I'm really happy with my purchase.
    I've been using the product for two weeks now.
    It does exactly as described in the product description.
    The only problem is that it takes a long time to charge.
    However, since I recharge during nights, this is something I can
    live with.
    """

sentence_list = tokenize.sent_tokenize(paragraph)
cumulative_sentiment = 0.0
for sentence in sentence_list:
    vs = analyzer.polarity_scores(sentence)
    cumulative_sentiment += vs["compound"]
    print(sentence, ' : ', vs["compound"])

average_sentiment = cumulative_sentiment / len(sentence_list)
print('Average score:', average_score)
```

If you run this code, ntlk will most likely complain about the fact that a resource is missing:

```
Resource punkt not found.
  Please use the NLTK Downloader to obtain the resource:

  >>> import nltk
  >>> nltk.download('punkt')
[...]
```

To fix this, execute the recommended commands on a Python shell:

```
>>> import nltk
>>> nltk.download('punkt')
```

After the resource has been downloaded and installed, the code above should work fine and will output:

```
I'm really happy with my purchase. : 0.6115
I've been using the product for two weeks now. : 0.0
It does exactly as described in the product description. : 0.0
The only problem is that it takes a long time to charge. : -0.4019
However, since I recharge during nights, this is something I can live
with. : 0.0
Average score: 0.04192000000000001
```

Let's apply this to our list of Amazon reviews. We'll calculate the sentiment for each rating, organize them by rating, and then use the "matplotlib" library to draw violin plots of the sentiment scores per rating:

```python
from vaderSentiment.vaderSentiment import SentimentIntensityAnalyzer
from nltk import tokenize
import dataset
import matplotlib.pyplot as plt

db = dataset.connect('sqlite:///reviews.db')
reviews = db['reviews'].all()

analyzer = SentimentIntensityAnalyzer()

sentiment_by_stars = [[] for r in range(1,6)]

for review in reviews:
    full_review = review['title'] + '. ' + review['review']
    sentence_list = tokenize.sent_tokenize(full_review)
    cumulative_sentiment = 0.0
    for sentence in sentence_list:
        vs = analyzer.polarity_scores(sentence)
        cumulative_sentiment += vs["compound"]
    average_score = cumulative_sentiment / len(sentence_list)
    sentiment_by_stars[int(review['rating'])-1].append(average_score)
```

```
plt.violinplot(sentiment_by_stars,
               range(1,6),
               vert=False, widths=0.9,
               showmeans=False, showextrema=True, showmedians=True,
               bw_method='silverman')
plt.axvline(x=0, linewidth=1, color='black')
plt.show()
```

This should output a figure similar to the one shown in Figure 9-10. In this case, we can indeed observe a strong correlation between the rating and the sentiments of the texts, though it's interesting to note that even for lower ratings (two and three stars), the majority of reviews are still somewhat positive. Of course, there is a lot more that can be done with this data set. Think, for instance, about a predictive model to detect fake reviews.

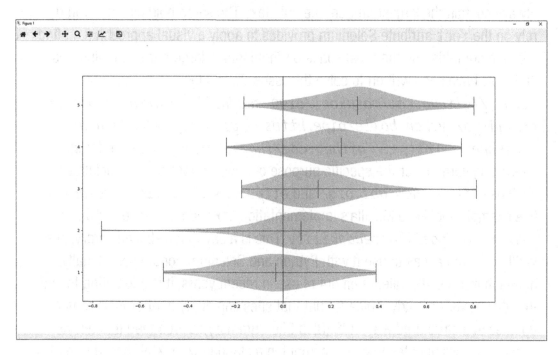

Figure 9-10. *Sentiment plots per rating level*

9.12 Scraping and Analyzing News Articles

We're going to use Selenium to scrape the "Top Stories" from Google News, see *https:// news.google.com/news/?ned=us&hl=en*. Our goal is to visit every article and get out the title and main content of the article.

Not as Easy as It Looks   Getting out the "main content" from a page is trickier as it might seem at first sight. You might try to iterate all the lowest-level HTML elements and keeping the one with the most text embedded in it, though this approach will break if the text in an article is split up over multiple sibling elements, like a series of "<p>" tags inside a larger "<div>", for instance. Considering all elements does not resolve this issue, as you'll end up by simply selecting the top element (e.g., "<html>" or "<body>") on the page, as this will always contain the largest amount (i.e., all) text. The same holds in case you'd rely on the `rect` attribute Selenium provides to apply a visual approach (i.e., find the element taking up the most space on the page). A large number of libraries and tools have been written to solve this issue. Take a look at, for example, *https://github.com/masukomi/ar90-readability*, *https://github. com/misja/python-boilerpipe*, *https://github.com/codelucas/ newspaper* and *https://github.com/fhamborg/news-please* for some interesting libraries for the specific purpose of news extraction, or specialized APIs such as *https://newsapi.org/* and *https://webhose.io/news-api*. In this example, we'll use Mozilla's implementation of Readability; see *https:// github.com/mozilla/readability*. This is a JavaScript-based library, but we'll figure out a way to use it with Python and Selenium nonetheless. Finally, although it has sadly fallen a bit out of use in recent years, it is interesting to know that there exists already a nice format that sites can apply to offer their content updates in a structured way: RSS (Rich Site Summary): a web feed that allows users to access updates to online content in a standardized, XML-based format. Keep an eye out for "<link>" tags with their "type" attribute set to "application/ rss+xml". The "href" attribute will then announce the URL where the RSS feed can be found.

Let's start by getting out a list of "Top Stories" links from Google News using Selenium. A first iteration of our script looks as follows:

```
from selenium import webdriver

base_url = 'https://news.google.com/news/?ned=us&hl=en'

driver = webdriver.Chrome()
driver.implicitly_wait(10)
driver.get(base_url)

for link in driver.find_elements_by_css_selector('main a[role="heading"]'):
    news_url = link.get_attribute('href')
    print(news_url)

driver.quit()
```

This will output the following (of course, your links might vary):

```
http://news.xinhuanet.com/english/2017-10-24/c_136702615.htm
http://www.cnn.com/2017/10/24/asia/china-xi-jinping-thought/index.html
[...]
```

Navigate to *http://edition.cnn.com/2017/10/24/asia/china-xi-jinping-thought/index.html* in your browser and open your browser's console in its developer tools. Our goal is now to extract the content from this page, using Mozilla's Readability implementation in JavaScript, a tool which is normally used to display articles in a more readable format. That is, we would like to "inject" the JavaScript code available at *https://raw.githubusercontent.com/mozilla/readability/master/Readability.js* in the page. Since we are able to instruct the browser to execute JavaScript using Selenium, we hence need to come up with an appropriate piece of JavaScript code to perform this injection. Using your browser's console, try executing the following block of code:

```
(function(d, script) {
  script = d.createElement('script');
  script.type = 'text/javascript';
  script.async = true;
  script.onload = function(){
    console.log('The script was successfully injected!');
  };
```

```
script.src = 'https://raw.githubusercontent.com/' +
      'mozilla/readability/master/Readability.js';
d.getElementsByTagName('head')[0].appendChild(script);
}(document));
```

This script works as follows: a new "<script>" element is constructed with its "src" parameter set to *https://raw.githubusercontent.com/mozilla/readability/master/Readability.js*, and appended into the "<head>" of the document. Once the script has loaded, we show a message on the console. This will provide you with a result as shown in Figure 9-11.

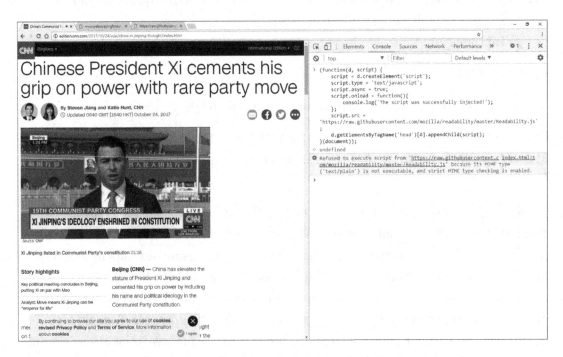

Figure 9-11. *Trying to inject a "<script>" tag using JavaScript*

This does not work as we had expected, as Chrome refuses to execute this script:

```
Refused to execute script from
'https://raw.githubusercontent.com/mozilla/readability/master/Readability.js'
because its MIME type ('text/plain') is not executable, and strict MIME ↵
type checking  is enabled.
```

The problem here is that GitHub indicates in its headers that the content type of this document is "text/plain," and Chrome prevents us from using it as a script. To work around this issue, we'll host a copy of the script ourselves at *http://www. webscrapingfordatascience.com/readability/Readability.js* and try again:

```
(function(d, script) {
  script = d.createElement('script');
  script.type = 'text/javascript';
  script.async = true;
  script.onload = function(){
    console.log('The script was successfully injected!');
  };
  script.src = 'http://www.webscrapingfordatascience.com/readability/
  Readability.js';
  d.getElementsByTagName('head')[0].appendChild(script);
}(document));
```

Which should give the correct result:

```
The script was successfully injected!
```

Now that the "<script>" tag has been injected and executed, but we need to figure out how to use it. Mozilla's documentation at *https://github.com/mozilla/ readability* provides us with some instructions, based on which we can try executing the following (still in the console window):

```
var documentClone = document.cloneNode(true);
var loc = document.location;
var uri = {
  spec: loc.href,
  host: loc.host,
  prePath: loc.protocol + "//" + loc.host,
  scheme: loc.protocol.substr(0, loc.protocol.indexOf(":")),
  pathBase: loc.protocol + "//" + loc.host +
            loc.pathname.substr(0, loc.pathname.lastIndexOf("/") + 1)
};

var article = new Readability(uri, documentClone).parse();
console.log(article);
```

This should provide you with a result as shown in Figure 9-12, which looks promising indeed: the "article" object contains a "title" and "content" attribute we'll be able to use.

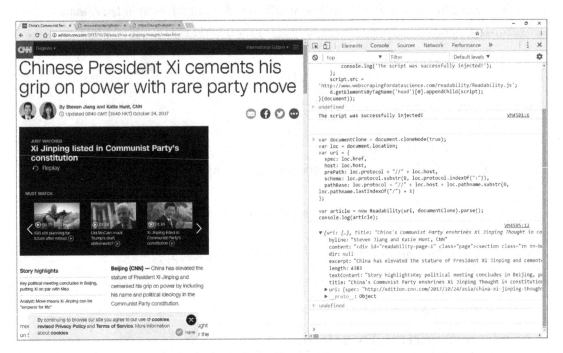

Figure 9-12. *Extracting the article's information*

The question is now how we can return this information to Selenium. Remember that we can execute JavaScript commands from Selenium through the `execute_script` method. One possible approach to get out the information we want is to use JavaScript to replace the whole page's contents with the information we want, and then use Selenium to get out that information:

```python
from selenium import webdriver

base_url = 'http://edition.cnn.com/2017/10/24/asia/china-xi-jinping-
thought/index.html'

driver = webdriver.Chrome()
driver.implicitly_wait(10)

driver.get(base_url)
```

```
js_cmd = '''
(function(d, script) {
  script = d.createElement('script');
  script.type = 'text/javascript';
  script.async = true;
  script.onload = function(){
    var documentClone = document.cloneNode(true);
    var loc = document.location;
    var uri = {
      spec: loc.href,
      host: loc.host,
      prePath: loc.protocol + "//" + loc.host,
      scheme: loc.protocol.substr(0, loc.protocol.indexOf(":")),
      pathBase: loc.protocol + "//" + loc.host +
                loc.pathname.substr(0, loc.pathname.lastIndexOf("/") + 1)
    };
    var article = new Readability(uri, documentClone).parse();
    document.body.innerHTML = '<h1 id="title">' + article.title + '</h1>' +
      '<div id="content">' + article.content + '</div>';
  };
  script.src = 'http://www.webscrapingfordatascience.com/readability/
  Readability.js';
d.getElementsByTagName('head')[0].appendChild(script);
}(document));
'''

driver.execute_script(js_cmd)

title = driver.find_element_by_id('title').text.strip()
content = driver.find_element_by_id('content').text.strip()

print('Title was:', title)

driver.quit()
```

The "document.body.innerHTML" line in the JavaScript command will replace the contents of the "<body>" tag with a header and a "<div>" tag, from which we can then simply retrieve our desired information.

However, the execute_script method also allows us to pass back JavaScript objects to Python, so the following approach also works:

```python
from selenium import webdriver
from selenium.webdriver.common.by import By
from selenium.webdriver.support.ui import WebDriverWait
from selenium.webdriver.support import expected_conditions as EC

base_url = 'http://edition.cnn.com/2017/10/24/asia/china-xi-jinping-
thought/index.html'

driver = webdriver.Chrome()
driver.implicitly_wait(10)

driver.get(base_url)

js_cmd = '''
(function(d, script) {
  script = d.createElement('script');
  script.type = 'text/javascript';
  script.async = true;
  script.onload = function() {
      script.id = 'readability-script';
  }
  script.src = 'http://www.webscrapingfordatascience.com/readability/
  Readability.js';
  d.getElementsByTagName('head')[0].appendChild(script);
}(document));
'''

js_cmd2 = '''
var documentClone = document.cloneNode(true);
var loc = document.location;
var uri = {
  spec: loc.href,
  host: loc.host,
  prePath: loc.protocol + "//" + loc.host,
  scheme: loc.protocol.substr(0, loc.protocol.indexOf(":")),
```

```
    pathBase: loc.protocol + "//" + loc.host +
            loc.pathname.substr(0, loc.pathname.lastIndexOf("/") + 1)
};
var article = new Readability(uri, documentClone).parse();
return JSON.stringify(article);
'''

driver.execute_script(js_cmd)

wait = WebDriverWait(driver, 10)
wait.until(EC.presence_of_element_located((By.ID, "readability-script")))

returned_result = driver.execute_script(js_cmd2)

print(returned_result)

driver.quit()
```

There are several intricacies here that warrant some extra information. First, note that we're using the execute_script method twice: once to inject the "<script>" tag, and then again to get out our "article" object. However, since executing the script might take some time, and Selenium's implicit wait does not take this into account when using execute_script, we use an explicit wait to check for the presence of an element with an "id" of "readability-script," which is set by the "script.onload" function. Once such an id is found, we know that the script has finished loading and we can execute the second JavaScript command. Here, we do need to use "JSON.stringify" to make sure we return a JSON-formatted string instead of a raw JavaScript object to Python, as Python will not be able to make sense of this return value and convert it to a list of None values (simple types, such as integers and strings, are fine, however).

Let's clean up our script a little and merge it with our basic framework:

```
from selenium import webdriver
from selenium.webdriver.common.by import By
from selenium.webdriver.support.ui import WebDriverWait
from selenium.webdriver.support import expected_conditions as EC

base_url = 'https://news.google.com/news/?ned=us&hl=en'
```

```
inject_readability_cmd = '''
(function(d, script) {
  script = d.createElement('script');
  script.type = 'text/javascript';
  script.async = true;
  script.onload = function() {
      script.id = 'readability-script';
  }
  script.src = 'http://www.webscrapingfordatascience.com/readability/
  Readability.js';
  d.getElementsByTagName('head')[0].appendChild(script);
}(document));
'''

get_article_cmd = '''
var documentClone = document.cloneNode(true);
var loc = document.location;
var uri = {
  spec: loc.href,
  host: loc.host,
  prePath: loc.protocol + "//" + loc.host,
  scheme: loc.protocol.substr(0, loc.protocol.indexOf(":")),
  pathBase: loc.protocol + "//" + loc.host +
            loc.pathname.substr(0, loc.pathname.lastIndexOf("/") + 1)
};
var article = new Readability(uri, documentClone).parse();
return JSON.stringify(article);
'''

driver = webdriver.Chrome()
driver.implicitly_wait(10)

driver.get(base_url)

news_urls = []
for link in driver.find_elements_by_css_selector('main a[role="heading"]'):
    news_url = link.get_attribute('href')
    news_urls.append(news_url)
```

```
for news_url in news_urls:
    print('Now scraping:', news_url)
    driver.get(news_url)

    print('Injecting scripts')
    driver.execute_script(inject_readability_cmd)
    wait = WebDriverWait(driver, 10)
    wait.until(EC.presence_of_element_located((By.ID, "readability-script")))
    returned_result = driver.execute_script(get_article_cmd)

    # Do something with returned_result

driver.quit()
```

Note that we're using two "for" loops: one to extract the links we wish to scrape, which we'll store in a list; and another one to iterate over the list. Using one loop wouldn't work in this case: as we're navigating to other pages inside of the loop, Selenium would complain about "stale elements" when trying to find the next link with find_elements_by_css_selector. This is basically saying: "I'm trying to find the next element for you, but the page has changed in the meantime, so I can't be sure anymore what you want to retrieve."

If you try to execute this script, you'll note that it quickly fails anyway. What is happening here? To figure out what is going wrong, try opening another link in your browser, say *https://www.washingtonpost.com/world/chinas-leader-elevated-to-the-level-of-mao-in-communist-pantheon/2017/10/24/ddd911e0-b832-11e7-9b93-b97043e57a22_story.html?utm_term=.720e06a5017d* (a site using HTTPS), and executing the first JavaScript command manually in your browser's console, that is, by copy-pasting and executing:

```
(function(d, script) {
  script = d.createElement('script');
  script.type = 'text/javascript';
  script.async = true;
  script.onload = function() {
      script.id = 'readability-script';
  }
}
```

```
script.src = 'http://www.webscrapingfordatascience.com/readability/
Readability.js';
  d.getElementsByTagName('head')[0].appendChild(script);
}(document));
```

You'll probably get a result like what follows:

```
GET https://www.webscrapingfordatascience.com/readability/Readability.js ↵
net::ERR_CONNECTION_CLOSED
```

On other pages, you might get:

```
Mixed Content: The page at [...] was loaded over HTTPS, but requested an ↵
insecure script 'http://www.webscrapingfordatascience.com/readability/
Readability.js'.
This request has been blocked; the content must be served over HTTPS.
```

It's clear what's going on here: if we load a website through HTTPS and try to inject a script through HTTP, Chrome will block this request as it deems it insecure (which is true). Other sites might apply other approaches to prevent script injection, using for example, a "Content-Security-Policy" header. that would result in an error like this:

```
Refused to load the script
    'http://www.webscrapingfordatascience.com/readability/Readability.js' ↵
    because it violates the following Content Security Policy directive: ↵
    "script-src 'self' 'unsafe-eval' 'unsafe-inline'".
```

There are extensions available for Chrome that will disable such checks, but we're going to take a different approach here, which will work on the majority of pages except those with the most strict Content Security Policies: instead of trying to inject a "<script>" tag, we're going to simply take the contents of our JavaScript file and execute these directly using Selenium. We can do so by loading the contents from a local file, but since we've already hosted the file online, we're going to use requests to fetch the contents instead:

```
from selenium import webdriver
import requests
```

```
base_url = 'https://news.google.com/news/?ned=us&hl=en'
script_url = 'http://www.webscrapingfordatascience.com/readability/
Readability.js'

get_article_cmd = requests.get(script_url).text
get_article_cmd += '''

var documentClone = document.cloneNode(true);
var loc = document.location;
var uri = {
  spec: loc.href,
  host: loc.host,
  prePath: loc.protocol + "//" + loc.host,
  scheme: loc.protocol.substr(0, loc.protocol.indexOf(":")),
  pathBase: loc.protocol + "//" + loc.host +
            loc.pathname.substr(0, loc.pathname.lastIndexOf("/") + 1)
};
var article = new Readability(uri, documentClone).parse();
return JSON.stringify(article);
'''

driver = webdriver.Chrome()
driver.implicitly_wait(10)

driver.get(base_url)

news_urls = []
for link in driver.find_elements_by_css_selector('main a[role="heading"]'):
    news_url = link.get_attribute('href')
    news_urls.append(news_url)

for news_url in news_urls:
    print('Now scraping:', news_url)
    driver.get(news_url)

    print('Injecting script')
    returned_result = driver.execute_script(get_article_cmd)

    # Do something with returned_result
driver.quit()
```

This approach also has the benefit that we can execute our whole JavaScript command in one go and do not need to rely on an explicit wait anymore to check whether the script has finished loading. The only thing remaining now is to convert the retrieved result to a Python dictionary and store our results in a database, once more using the "dataset" library:

```python
from selenium import webdriver
import requests
import dataset
from json import loads

db = dataset.connect('sqlite:///news.db')

base_url = 'https://news.google.com/news/?ned=us&hl=en'
script_url = 'http://www.webscrapingfordatascience.com/readability/
Readability.js'

get_article_cmd = requests.get(script_url).text
get_article_cmd += '''

var documentClone = document.cloneNode(true);
var loc = document.location;
var uri = {
  spec: loc.href,
  host: loc.host,
  prePath: loc.protocol + "//" + loc.host,
  scheme: loc.protocol.substr(0, loc.protocol.indexOf(":")),
  pathBase: loc.protocol + "//" + loc.host +
            loc.pathname.substr(0, loc.pathname.lastIndexOf("/") + 1)
};
var article = new Readability(uri, documentClone).parse();
return JSON.stringify(article);
'''

driver = webdriver.Chrome()
driver.implicitly_wait(10)

driver.get(base_url)
```

```
news_urls = []
for link in driver.find_elements_by_css_selector('main a[role="heading"]'):
    news_url = link.get_attribute('href')
    news_urls.append(news_url)

for news_url in news_urls:
    print('Now scraping:', news_url)
    driver.get(news_url)

    print('Injecting script')
    returned_result = driver.execute_script(get_article_cmd)

    # Convert JSON string to Python dictionary
    article = loads(returned_result)
    if not article:
        # Failed to extract article, just continue
        continue
    # Add in the url
    article['url'] = news_url
    # Remove 'uri' as this is a dictionary on its own
    del article['uri']
    # Add to the database
    db['articles'].upsert(article, ['url'])

    print('Title was:', article['title'])

driver.quit()
```

The output looks as follows:

```
Now scraping: https://www.usnews.com/news/world/articles/2017-10-24/
              china-southeast-asia-aim-to-build-trust-with-sea-drills-
              singapore-says Injecting script
Title was: China, Southeast Asia Aim to Build Trust With Sea Drills,     ↵
           Singapore Says | World News
```

```
Now scraping: http://www.philstar.com/headlines/2017/10/24/
            1751999/pentagon-chief-seeks-continued-maritime-cooperation-
            asean Injecting script
Title was: Pentagon chief seeks continued maritime cooperation with ASEAN |
            Headlines News, The Philippine Star,
[...]
```

Remember to take a look at the database ("news.db") using a SQLite client such as "DB Browser for SQLite"; see Figure 9-13.

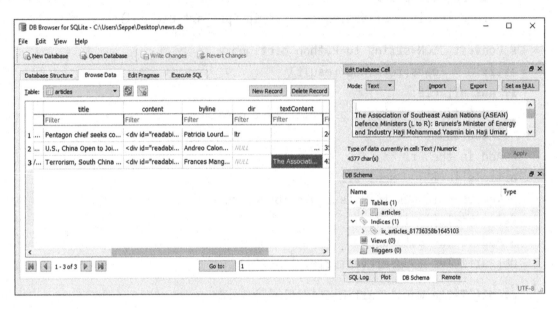

Figure 9-13. *Exploring some scraped articles with DB Browser for SQLite*

We can now analyze our collected articles using Python. We're going to construct a topic model using Latent Dirichlet Allocation (LDA) that will help us to categorize our articles along some topics. To do so, we'll use the "nltk," "stop-words," and "gensim" libraries, which can simply be installed using pip:

```
pip install -U nltk
pip install -U stop-words
pip install -U gensim
```

First, we're going to loop through all our articles in order to tokenize them (convert text into a list of word elements) using a simple regular expression, remove stop words, and apply stemming:

```python
import dataset
from nltk.tokenize import RegexpTokenizer
from nltk.stem.porter import PorterStemmer
from stop_words import get_stop_words

db = dataset.connect('sqlite:///news.db')

articles = []

tokenizer = RegexpTokenizer(r'\w+')
stop_words = get_stop_words('en')
p_stemmer = PorterStemmer()

for article in db['articles'].all():
    text = article['title'].lower().strip()
    text += " " + article['textContent'].lower().strip()
    if not text:
        continue
    # Tokenize
    tokens = tokenizer.tokenize(text)
    # Remove stop words and small words
    clean_tokens = [i for i in tokens if not i in stop_words]
    clean_tokens = [i for i in clean_tokens if len(i) > 2]
    # Stem tokens
    stemmed_tokens = [p_stemmer.stem(i) for i in clean_tokens]
    # Add to list
    articles.append((article['title'], stemmed_tokens))

print(articles[0])
```

Our first article now looks as follows (we keep the title for later reporting):

```
('Paul Manafort, former business partner to surrender in Mueller
investigation', ['presid', 'trump', 'former', 'campaign', 'chairman', [...]])
```

To generate an LDA model, we need to calculate how frequently each term occurs within each document. To do that, we can construct a document-term matrix with gensim:

```
from gensim import corpora

dictionary = corpora.Dictionary([a[1] for a in articles])
corpus = [dictionary.doc2bow(a[1]) for a in articles]

print(corpus[0])
```

The Dictionary class traverses texts and assigns a unique integer identifier to each unique token while also collecting word counts and relevant statistics. Next, our dictionary is converted to a bag of words corpus that results in a list of vectors equal to the number of documents. Each document vector is a series of "(id, count)" tuples:

```
[(0, 10), (1, 17), (2, 7), (3, 11), [...]]
```

We're now ready to construct an LDA model:

```
from gensim.models.ldamodel import LdaModel

nr_topics = 30
ldamodel = LdaModel(corpus, num_topics=nr_topics,
                    id2word=dictionary, passes=20)

print(ldamodel.print_topics())
```

This will show something like:

```
[(0, '0.027*"s" + 0.018*"trump" + 0.018*"manafort" + 0.011*"investig"    ↵
     + 0.008*"presid" + 0.008*"report" + 0.007*"mueller" + 0.007*"year"   ↵
     + 0.007*"campaign" + 0.006*"said"'),
 (1, '0.014*"s" + 0.014*"said" + 0.013*"percent" + 0.008*"1" +            ↵
     0.007*"0" + 0.006*"year" + 0.006*"month" + 0.005*"increas" +         ↵
     0.005*"3" + 0.005*"spend"'),
 [...]
]
```

This overview shows an entry per topic. Each topic is represented by a list of probable words to appear in that topic, ordered by probability of appearance. Note that adjusting the model's number and amount of "passes" is important to get a good result. Once the results look acceptable (we've increased the number of topics for our scraped set), we can use our model to assign topics to our documents:

```python
from random import shuffle

# Show topics by top-3 terms
for t in range(nr_topics):
    print(ldamodel.print_topic(t, topn=3))

# Show some random articles
idx = list(range(len(articles)))
shuffle(idx)
for a in idx[:3]:
    article = articles[a]
    print('===========================')
    print(article[0])
    prediction = ldamodel[corpus[a]][0]
    print(ldamodel.print_topic(prediction[0], topn=3))
    print('Probability:', prediction[1])
```

This will show something like the following:

```
0.014*"new" + 0.013*"power" + 0.013*"storm"
0.030*"rapp" + 0.020*"spacey" + 0.016*"said"
0.024*"catalan" + 0.020*"independ" + 0.019*"govern"
0.025*"manafort" + 0.020*"trump" + 0.015*"investig"
0.007*"quickli" + 0.007*"complex" + 0.007*"deal"
0.018*"earbud" + 0.016*"iconx" + 0.014*"samsung"
0.012*"halloween" + 0.007*"new" + 0.007*"star"
0.021*"octopus" + 0.014*"carver" + 0.013*"vega"
0.000*"rapp" + 0.000*"spacey" + 0.000*"said"
0.025*"said" + 0.017*"appel" + 0.012*"storm"
0.039*"akzo" + 0.018*"axalta" + 0.017*"billion"
```

```
0.024*"rapp" + 0.024*"spacey" + 0.017*"said"
0.000*"boehner" + 0.000*"one" + 0.000*"trump"
0.033*"boehner" + 0.010*"say" + 0.009*"hous"
0.000*"approv" + 0.000*"boehner" + 0.000*"quarter"
0.017*"tax" + 0.013*"republican" + 0.011*"week"
0.012*"trump" + 0.008*"plan" + 0.007*"will"
0.005*"ludwig" + 0.005*"underlin" + 0.005*"sensibl"
0.015*"tax" + 0.011*"trump" + 0.011*"look"
0.043*"minist" + 0.032*"prime" + 0.030*"alleg"
0.058*"harri" + 0.040*"polic" + 0.032*"old"
0.040*"musk" + 0.026*"tunnel" + 0.017*"compani"
0.055*"appl" + 0.038*"video" + 0.027*"peterson"
0.011*"serv" + 0.008*"almost" + 0.007*"insid"
0.041*"percent" + 0.011*"year" + 0.010*"trump"
0.036*"univers" + 0.025*"econom" + 0.012*"special"
0.022*"chees" + 0.021*"patti" + 0.019*"lettuc"
0.000*"boehner" + 0.000*"said" + 0.000*"year"
0.000*"boehner" + 0.000*"new" + 0.000*"say"
0.030*"approv" + 0.025*"quarter" + 0.021*"rate"
==========================
Paul Manafort, Who Once Ran Trump Campaign, Indicted on Money Laundering
and Tax Charges
0.025*"manafort" + 0.020*"trump" + 0.015*"investig"
Probability: 0.672658189483

==========================
Apple fires employee after daughter's iPhone X video goes viral
0.055*"appl" + 0.038*"video" + 0.027*"peterson"
Probability: 0.990880503145

==========================
Theresa May won't say when she knew about sexual harassment allegations
0.043*"minist" + 0.032*"prime" + 0.030*"alleg"
Probability: 0.774530402797
```

Scraping Topics   There is still a lot of room to improve on this by, for example, exploring other topic model mapping algorithms, applying better tokenization, adding custom stop words, or expanding the set of articles or adjusting the parameters. Alternatively, you might also consider scraping the tags for each article straight from the Google News page, which also includes these as "topics" on its page.

9.13 Scraping and Analyzing a Wikipedia Graph

In this example, we'll work once again with Wikipedia (we already used Wikipedia in the chapter on web crawling). Our goal here is to scrape titles of Wikipedia pages, while keeping track of links between them, which we'll use to construct a graph and analyze it using Python. We'll again use the "dataset" library as a simple means to store results. The following code contains the full crawling setup:

```python
import requests
import dataset
from bs4 import BeautifulSoup
from urllib.parse import urljoin, urldefrag
from joblib import Parallel, delayed

db = dataset.connect('sqlite:///wikipedia.db')
base_url = 'https://en.wikipedia.org/wiki/'

def store_page(url, title):
    print('Visited page:', url)
    print(' title:', title)
    db['pages'].upsert({'url': url, 'title': title}, ['url'])

def store_links(from_url, links):
    db.begin()
    for to_url in links:
        db['links'].upsert({'from_url': from_url, 'to_url': to_url},
                           ['from_url', 'to_url'])
    db.commit()
```

```python
def get_random_unvisited_pages(amount=10):
    result = db.query('''SELECT * FROM links
        WHERE to_url NOT IN (SELECT url FROM pages)
        ORDER BY RANDOM() LIMIT {}'''.format(amount))
    return [r['to_url'] for r in result]

def should_visit(base_url, url):
    if url is None:
        return None
    full_url = urljoin(base_url, url)
    full_url = urldefrag(full_url)[0]
    if not full_url.startswith(base_url):
        # This is an external URL
        return None
    ignore = ['Wikipedia:', 'Template:', 'File:', 'Talk:', 'Special:',
            'Template talk:', 'Portal:', 'Help:', 'Category:', 'index.php']
    if any([i in full_url for i in ignore]):
        # This is a page to be ignored
        return None
    return full_url

def get_title_and_links(base_url, url):
    html = requests.get(url).text
    html_soup = BeautifulSoup(html, 'html.parser')
    page_title = html_soup.find(id='firstHeading')
    page_title = page_title.text if page_title else ''
    links = []
    for link in html_soup.find_all("a"):
        link_url = should_visit(base_url, link.get('href'))
        if link_url:
            links.append(link_url)
    return url, page_title, links

if __name__ == '__main__':
    urls_to_visit = [base_url]
```

```
while urls_to_visit:
    scraped_results = Parallel(n_jobs=5, backend="threading")(
        delayed(get_title_and_links)(base_url, url) for url in
        urls_to_visit
    )
    for url, page_title, links in scraped_results:
        store_page(url, page_title)
        store_links(url, links)
    urls_to_visit = get_random_unvisited_pages()
```

There are a lot of things going on here that warrant some extra explanation:

- The database is structured as follows: a table "pages" holds a list of visited URLs with their page titles. The method store_page is used to store entries in this table. Another table, "links," simply contains pairs of URLs to represent links between pages. The method store_link is used to update these, and both methods use the "dataset" library. For the latter, we perform multiple upsert operations inside a single explicit database transaction to speed things up.

- The method get_random_unvisited_pages now returns a list of unvisited URLs, rather than just one, by selecting a random list of linked-to URLs that do not yet appear in the "pages" table (and hence have not been visited yet).

- The should_visit method is used to determine whether a link should be considered for crawling. It returns a proper formatted URL if it should be included, or None otherwise.

- The get_title_and_links method performs the actual scraping of pages, fetching their title and a list of URLs.

- The script itself loops until there are no more unvisited pages (basically forever, as new pages will continue to be discovered). It fetches out a list of random pages we haven't visited yet, gets their title and links, and stores these in the database.

- Note that we use the "joblib" library here to set up a parallel approach. Simply visiting URLs one by one would be a tad too slow here, so we use joblib to set up a multithreaded approach to visit links at the same time, effectively spawning multiple network requests. It's important not to hammer our own connection or Wikipedia, so we limit the n_jobs argument to five. The back-end argument is used here to indicate that we want to set up a parallel calculation using multiple threads, instead of multiple processes. Both approaches have their pros and cons in Python. A multi-process approach comes with a bit more overhead to set up, but it can be faster as Python's internal threading system can be a bit tedious due to the "global interpreter lock" (the GIL) (a full discussion about the GIL is out of scope here, but feel free to look up more information online if this is the first time you have heard about it). In our case, the work itself is relatively straightforward: execute a network request and perform some parsing, so a multithreading approach is fine.

- This is also the reason why we don't store the results in the database inside the get_title_and_links method itself, but wait until the parallel jobs have finished their execution and have returned their results. SQLite doesn't like to be written to from multiple threads or many processes at once, so we wait until we have collected the results before writing them to the database. An alternative would be to use a client-server database system. Note that we should avoid overloading the database too much with a huge set of results. Not only will the intermediate results have to be stored in memory, but we'll also incur a waiting time when writing the large set of results. Since the get_random_unvisited_pages method returns a list of ten URLs maximum, we don't need to worry about this too much in our case.

- Finally, note that the main entry point of the script is now placed under "if __name__ == '__main__':". In other examples, we have not done so for the sake of simplicity, although it is good practice to do so nonetheless. The reason for this is as follows: when a Python script imports another module, all the code contained in that module is executed at once. For instance, if we'd like to reuse the should_visit method in another script, we could import our original script using "import myscript" or "from myscript import should_visit." In both cases, the full code in "myscript.py" will be executed. If this script contains a block of code, like our "while" loop in this example, it will start executing that block of code, which is not what we want when importing our script; we just want to load the function definitions. We hence want to indicate to Python to "only execute this block of code when the script is directly executed," which is what the "if __name__ == '__main__':" check does. If we start our script from the command line, the special "__name__" variable will be set to "__main__". If our script would be imported from another module, "__name__" will be set to that module's name instead. When using joblib as we do here, the contents of our script will be sent to all "workers" (threads or processes), in order for them to perform the correct imports and load the correct function definitions. In our case, for instance, the different workers should know about the get_title_ and_links method. However, since the workers will also execute the full code contained in the script (just like an import would), we also need to prevent them from running the main block of code as well, which is why we need to provide an "if __name__ == '__main__':" check.

You can let the crawler run for as long as you like, though note that it is extremely unlikely to ever finish, and a smaller graph will also be a bit easier to look at in the next step. Once it has run for a bit, simply interrupt it to stop it. Since we use "upsert," feel free to resume it later on (it will just continue to crawl based on where it left off).

We can now perform some fun graph analysis using the scraped results. In Python, there are two popular libraries available to do so, NetworkX (the "networkx" library in pip) and iGraph ("python-igraph" in pip). We'll use NetworkX here, as well as "matplotlib" to visualize the graph.

Graph Visualization Is Hard   As the NetworkX documentation itself notes, proper graph visualization is hard, and the library authors recommend that people visualize their graphs with tools dedicated to that task. For our simple use case, the built-in methods suffice, even although we'll have to wrangle our way though matplotlib to make things a bit more appealing. Take a look at programs such as Cytoscape, Gephi, and Graphviz if you're interested in graph visualization. In the next example, we'll use Gephi to handle the visualization workload.

The following code visualizes the graph. We first construct a new NetworkX graph object and add in the pages as visited nodes. Next, we add the edges, though only between pages that were visited. As an extra step, we also remove nodes that are completely unconnected (even although these should not be present at this stage). We then calculate a centrality measure, called betweenness, as a measure of importance of nodes. This metric is calculated based on the number of shortest paths from all nodes to all other nodes that pass through the node we're calculating the metric for. The more times a node lies on the shortest path between two other nodes, the more important it is according to this metric. We'll color the nodes based on this metric by giving them different shades of blue. We apply a quick and dirty sigmoid function to the betweenness metric to "squash" the values in a range that will result in a more appealing visualization. We also add labels to nodes manually here, in order to have them appear above the actual nodes. This will provide a result as shown in Figure 9-14.

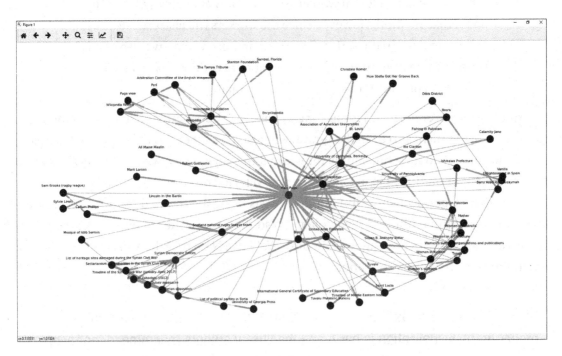

Figure 9-14. *Visualizing our scraped graph*

Ignore the Warnings   When running the visualization code, you'll most likely see warnings appear from matplotlib complaining about the fact that NetworkX is using deprecated functions. This is fine and can be safely ignored, though future versions of matplotlib might not play nice with NetworkX anymore. It's unclear whether the authors of NetworkX will continue to focus on visualization in the future. As you'll note, the "arrows" of the edges in the visualization also don't look very pretty. This is a long-standing issue with NetworkX. Again: NetworkX is fine for analysis and graph wrangling, though less so for visualization. Take a look at other libraries if visualization is your core concern.

```python
import networkx
import matplotlib.pyplot as plt
import dataset

db = dataset.connect('sqlite:///wikipedia.db')
G = networkx.DiGraph()

print('Building graph...')
for page in db['pages'].all():
    G.add_node(page['url'], title=page['title'])

for link in db['links'].all():
    # Only addedge if the endpoints have both been visited
    if G.has_node(link['from_url']) and G.has_node(link['to_url']):
        G.add_edge(link['from_url'], link['to_url'])

# Unclutter by removing unconnected nodes
G.remove_nodes_from(networkx.isolates(G))

# Calculate node betweenness centrality as a measure of importance
print('Calculating betweenness...')
betweenness = networkx.betweenness_centrality(G, endpoints=False)

print('Drawing graph...')

# Sigmoid function to make the colors (a little) more appealing
squash = lambda x : 1 / (1 + 0.5**(20*(x-0.1)))
```

```
colors = [(0, 0, squash(betweenness[n])) for n in G.nodes()]
labels = dict((n, d['title']) for n, d in G.nodes(data=True))
positions = networkx.spring_layout(G)

networkx.draw(G, positions, node_color=colors, edge_color='#AEAEAE')

# Draw the labels manually to make them appear above the nodes
for k, v in positions.items():

    plt.text(v[0], v[1]+0.025, s=labels[k],
             horizontalalignment='center', size=8)

plt.show()
```

9.14 Scraping and Visualizing a Board Members Graph

In this example, our goal is to construct a social graph of S&P 500 companies and their interconnectedness through their board members. We'll start from the S&P 500 page at Reuters available at *https://www.reuters.com/finance/markets/index/.SPX* to obtain a list of stock symbols:

```
from bs4 import BeautifulSoup
import requests
import re

session = requests.Session()

sp500 = 'https://www.reuters.com/finance/markets/index/.SPX'

page = 1
regex = re.compile(r'\/finance\/stocks\/overview\/.*')
symbols = []

while True:
    print('Scraping page:', page)
    params = params={'sortBy': '', 'sortDir' :'', 'pn': page}
    html = session.get(sp500, params=params).text
    soup = BeautifulSoup(html, "html.parser")
```

```
    pagenav = soup.find(class_='pageNavigation')
    if not pagenav:
        break
    companies = pagenav.find_next('table', class_='dataTable')
    for link in companies.find_all('a', href=regex):
        symbols.append(link.get('href').split('/')[-1])
    page += 1

print(symbols)
```

Once we have obtained a list of symbols, we can scrape the board member pages for each of them (e.g., *https://www.reuters.com/finance/stocks/company-officers/MMM.N*), fetch the table of board members, and store it as a pandas data frame, which we'll save using pandas' to_pickle method. Don't forget to install pandas first if you haven't already:

```
pip install -U pandas
```

Add this to the bottom of your script:

```
import pandas as pd

officers = 'https://www.reuters.com/finance/stocks/company-officers/{symbol}'

dfs = []

for symbol in symbols:
    print('Scraping symbol:', symbol)
    html = session.get(officers.format(symbol=symbol)).text
    soup = BeautifulSoup(html, "html.parser")
    officer_table = soup.find('table', {"class" : "dataTable"})
    df = pd.read_html(str(officer_table), header=0)[0]
    df.insert(0, 'symbol', symbol)
    dfs.append(df)

# Store the results
df = pd.concat(dfs)
df.to_pickle('sp500.pkl')
```

This sort of information can lead to a lot of interesting use cases, especially —
again — in the realm of graph and social network analytics. We're going to use NetworkX
once more, but simply to parse through our collected information and export a graph in
a format that can be read with Gephi, a popular graph visualization tool, which can be
downloaded from *https://gephi.org/users/download/*:

```python
import pandas as pd
import networkx as nx
from networkx.readwrite.gexf import write_gexf

df = pd.read_pickle('sp500.pkl')

G = nx.Graph()

for row in df.itertuples():
    G.add_node(row.symbol, type='company')
    G.add_node(row.Name, type='officer')
    G.add_edge(row.symbol, row.Name)

write_gexf(G, 'graph.gexf')
```

Open the graph file in Gephi, and apply the "ForceAtlas 2" layout technique for a
few iterations. We can also show labels as well, resulting in a figure like the one shown in
Figure 9-15.

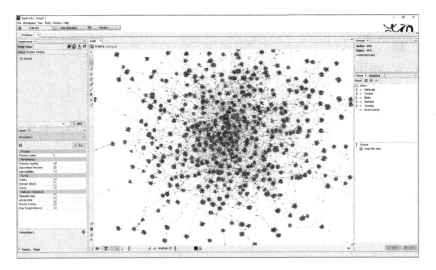

Figure 9-15. *Visualizing our scraped graph using Gephi*

Take some time to explore Gephi's visualization and filtering options if you like. All attributes that you have set in NetworkX ("type," in our case) will be available in Gephi as well. Figure 9-16 shows the filtered graph for Google, Amazon, and Apple with their board members, which are acting as connectors to other firms.

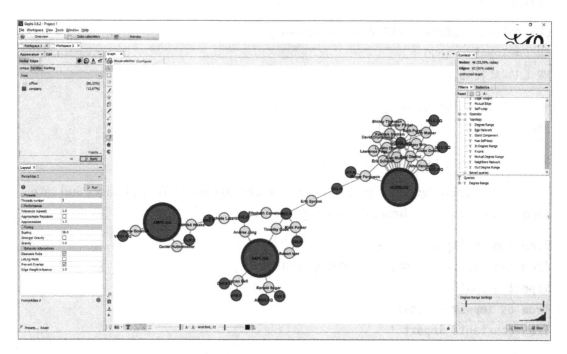

Figure 9-16. *Showing connected board members for Google, Amazon, and Apple*

9.15 Breaking CAPTCHA's Using Deep Learning

This final example is definitely the most challenging one, as well as the one that is mostly related to "data science," rather than web scraping. In fact, we'll not use any web scraping tools here. Instead, we're going to walk through a relatively contained example to illustrate how you could incorporate a predictive model in your web scraping pipeline in order to bypass a CAPTCHA check.

We're going to need to install some tools first. We'll use "OpenCV," an extremely thorough library for computer vision, as well as "numpy" for some basic data wrangling. Finally, we'll use the "captcha" library to generate example images. All of these can be installed as follows:

```
pip install -U opencv-python
pip install -U numpy
pip install -U captcha
```

Next, create a directory somewhere in your system to contain the Python scripts we will create. The first script ("constants.py") will contain some constants we're going to use:

```python
CAPTCHA_FOLDER = 'generated_images'
LETTERS_FOLDER = 'letters'

CHARACTERS = list('QWERTPASDFGHKLZXBNM')
NR_CAPTCHAS = 1000
NR_CHARACTERS = 4

MODEL_FILE = 'model.hdf5'
LABELS_FILE = 'labels.dat'

MODEL_SHAPE = (100, 100)
```

Another script ("generate.py") will generate a bunch of CAPTCHA images and save them to the "generated_images" directory:

```python
from random import choice
from captcha.image import ImageCaptcha
import os.path
from os import makedirs
from constants import *

makedirs(CAPTCHA_FOLDER)

image = ImageCaptcha()

for i in range(NR_CAPTCHAS):
    captcha = ''.join([choice(CHARACTERS) for c in range(NR_CHARACTERS)])
    filename = os.path.join(CAPTCHA_FOLDER, '{}_{}.png'.format(captcha, i))
    image.write(captcha, filename)
    print('Generated:', captcha)
```

After running this script, you should end up with a collection of CAPTCHA images (with their answers in the file names) as shown in Figure 9-17.

Figure 9-17. *A collection of generated CAPTCHA images*

Isn't This Cheating?   Of course, we're lucky here that we are generating the CAPTCHA's ourselves and hence have the opportunity to keep the answers as well. In the real world, however, CAPTCHA's do not expose their answer (it would kind of refute the point of the CAPTCHA), so that we would need to figure out another way to create our training set. One way is to look for the library a particular site is using to generate its CAPTCHA's and use it to collect a set of training images of your own, replicating the originals as closely as possible. Another approach is to manually label the images yourself, which is as dreadful as it sounds, though you might not need to label thousands of images to get the desired result. Since people make mistakes when filling in CAPTCHA's, too, we have more than one chance to get the answer right and hence do not need to target a 100 percent accuracy level. Even if our predictive model is only able to get one out of ten images right, that is still sufficient to break through a CAPTCHA after some retries.

Next, we're going to write another script that will cut up our images into separate pieces, one per character. We could try to construct a model that predicts the complete answer all at once, though in many cases it is much easier to perform the predictions character by character. To cut up our image, we'll need to invoke OpenCV to perform some heavy lifting for us. A complete discussion regarding OpenCV and computer vision would require a book in itself, so we'll stick to some basics here. The main concepts we'll use here are thresholding, opening, and contour detection. To see how this works, let's create a small test script first to show these concepts in action:

```python
import cv2
import numpy as np

# Change this to one of your generated images:
image_file = 'generated_images/ABQM_116.png'

image = cv2.imread(image_file)
cv2.imshow('Original image', image)

# Convert to grayscale, followed by thresholding to black and white
gray = cv2.cvtColor(image, cv2.COLOR_BGR2GRAY)
_, thresh = cv2.threshold(gray, 0, 255, cv2.THRESH_BINARY_INV | cv2.THRESH_OTSU)
cv2.imshow('Black and white', thresh)

# Apply opening: "erosion" followed by "dilation"
denoised = thresh.copy()
kernel = np.ones((4, 3), np.uint8)
denoised = cv2.erode(denoised, kernel, iterations=1)
kernel = np.ones((6, 3), np.uint8)
denoised = cv2.dilate(denoised, kernel, iterations=1)
cv2.imshow('Denoised', denoised)

# Now find contours and overlay them over our original image
_, cnts, _ = cv2.findContours(denoised.copy(), cv2.RETR_TREE, cv2.CHAIN_APPROX_NONE)
cv2.drawContours(image, cnts, contourIdx=-1, color=(255, 0, 0), thickness=-1)
cv2.imshow('Contours', image)

cv2.waitKey(0)
```

If you run this script, you should obtain a list of preview windows similar as shown in Figure 9-18. In the first two steps, we open our image with OpenCV and convert it to a simple pure black and white representation. Next, we apply an "opening" morphological transformation, which boils down to an erosion followed by dilation. The basic idea of erosion is just like soil erosion: this transformation "erodes away" boundaries of the foreground object (which is assumed to be in white) by sliding a "kernel" over the image (a "window," so to speak) so that only those white pixels are retained if all pixels in the surrounding kernel are white as well. Otherwise, it gets turned to black. Dilation does the opposite: it widens the image by setting pixels to white if at least one pixel in the surrounding kernel was white. Applying these steps is a very common tactic to remove noise from images. The kernel sizes used in the script above are simply the result of some trial and error, and you might want to adjust these with other types of CAPTCHA images. Note that we allow for some noise in the image to remain present. We don't need to obtain a perfect image as we trust that our predictive model will be able to "look over these."

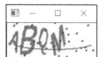

Figure 9-18. *Processing an image with OpenCV. From left to right: original image, image after conversion to black and white, image after applying an opening to remove noise, and the extracted contours overlaid in blue over the original image.*

Next, we use OpenCV's `findContours` method to extract "blobs" of connected white pixels. OpenCV comes with various methods to perform this extraction and different ways to represent the result (e.g., simplifying the contours or not, constructing a hierarchy or not, and so on). Finally, we use the `drawContours` method to draw the discovered blobs. The `contourIdx` argument here indicates that we want to draw all top-level contours, and the `thickness` value of -1 instructs OpenCV to fill up the contours.

We now still need a way to use the contours to create separate images: one per character. The way how we'll do so is by using masking. Note that OpenCV also allows to fetch out the "bounding rectangle" for each contour, which would make "cutting" the image much easier, though this might get us into trouble in case parts of the characters are near to each other. Instead, we'll use the approach illustrated by the following code fragment:

```python
import cv2
import numpy as np

image_file = 'generated_images/ABQM_116.png'

# Perform thresholding, erosion and contour finding as shown before
image = cv2.imread(image_file)
gray = cv2.cvtColor(image, cv2.COLOR_BGR2GRAY)
_, thresh = cv2.threshold(gray, 0, 255, cv2.THRESH_BINARY_INV | cv2.THRESH_
OTSU)
denoised = thresh.copy()
kernel = np.ones((4, 3), np.uint8)
denoised = cv2.erode(denoised, kernel, iterations=1)
kernel = np.ones((6, 3), np.uint8)
denoised = cv2.dilate(denoised, kernel, iterations=1)
_, cnts, _ = cv2.findContours(denoised.copy(), cv2.RETR_TREE, cv2.CHAIN_
APPROX_NONE)

# Create a fresh 'mask' image
mask = np.ones((image.shape[0], image.shape[1]), dtype="uint8") * 0
# We'll use the first contour as an example
contour = cnts[0]
# Draw this contour over the mask
cv2.drawContours(mask, [contour], -1, (255, 255, 255), -1)

cv2.imshow('Denoised image', denoised)

cv2.imshow('Mask after drawing contour', mask)

result = cv2.bitwise_and(denoised, mask)

cv2.imshow('Result after and operation', result)

retain = result > 0
result = result[np.ix_(retain.any(1), retain.any(0))]

cv2.imshow('Final result', result)

cv2.waitKey(0)
```

If you run this script, you'll obtain a result as shown in Figure 9-19. First, we create new black image with the same size as the starting, denoised image. We take one contour and draw it in white on top of this "mask." Next, the denoised image and mask are combined in a bitwise "and" operation, which will retain white pixels if the corresponding pixels in both input images were white, and sets it to black otherwise. Next, we apply some clever numpy slicing to crop the image.

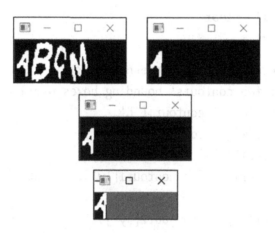

Figure 9-19. *Extracting part of an image using a contour mask in OpenCV. On the top left, the starting image is shown. On the right, a new image is created with the contour drawn in white and filled. These two images are combined in a bitwise "and" operation to obtain the image in the second row. The bottom image shows the final result after applying cropping.*

This is sufficient to get started, though there still is one problem we need to solve: overlap. In case characters overlap, they would be discovered as one large contour. To work around this issue, we'll apply the following operations. First, starting from a list of contours, check whether there is a significant degree of overlap between two distinct contours, in which case we only retain the largest one. Next, we order the contours based on their size, take the first n contours, and order these on the horizontal axis, from left to right (with n being the number of characters in a CAPTCHA). This still might lead to fewer contours than we need, so that we iterate over each contour, and check whether its width is higher than an expected value. A good heuristic for the expected value is to take the estimated width based on the distance from the leftmost white pixel to the rightmost white pixel divided by the number of characters we expect to see. In case a contour is wider than we expect, we cut it up into m equal parts, with m being equal to the width of the contour divided by the expected width. This is a heuristic that still might lead to some characters not being perfectly cut off (some characters are larger than others), but this is something we'll just accept. In case we don't end up with the desired number of characters at the end of all this, we'll simply skip over the given image.

We'll put all of this in a separate list of functions (in a file "functions.py"):

```python
import cv2
import numpy as np
from math import ceil, floor
from constants import *

def overlaps(contour1, contour2, threshold=0.8):
    # Check whether two contours' bounding boxes overlap
    area1 = contour1['w'] * contour1['h']
    area2 = contour2['w'] * contour2['h']
    left = max(contour1['x'], contour2['x'])
    right = min(contour1['x'] + contour1['w'], contour2['x'] +
    contour2['w'])
    top = max(contour1['y'], contour2['y'])
    bottom = min(contour1['y'] + contour1['h'], contour2['y'] +
    contour2['h'])
    if left <= right and bottom >= top:
        intArea = (right - left) * (bottom - top)
        intRatio = intArea / min(area1, area2)
        if intRatio >= threshold:
            # Return True if the second contour is larger
            return area2 > area1
    # Don't overlap or doesn't exceed threshold
    return None

def remove_overlaps(cnts):
    contours = []
    for c in cnts:
        x, y, w, h = cv2.boundingRect(c)
        new_contour = {'x': x, 'y': y, 'w': w, 'h': h, 'c': c}
        for other_contour in contours:
            overlap = overlaps(other_contour, new_contour)
```

```
            if overlap is not None:
                if overlap:
                    # Keep this one...
                    contours.remove(other_contour)
                    contours.append(new_contour)
                # ... otherwise do nothing: keep the original one
                break
        else:
            # We didn't break, so no overlap found, add the contour
            contours.append(new_contour)
    return contours

def process_image(image):
    # Perform basic pre-processing
    gray = cv2.cvtColor(image, cv2.COLOR_BGR2GRAY)
    _, thresh = cv2.threshold(gray, 0, 255, cv2.THRESH_BINARY_INV |
    cv2.THRESH_OTSU)
    denoised = thresh.copy()
    kernel = np.ones((4, 3), np.uint8)
    denoised = cv2.erode(denoised, kernel, iterations=1)
    kernel = np.ones((6, 3), np.uint8)
    denoised = cv2.dilate(denoised, kernel, iterations=1)
    return denoised

def get_contours(image):
    # Retrieve contours
    _, cnts, _ = cv2.findContours(image.copy(), cv2.RETR_TREE, cv2.CHAIN_
    APPROX_NONE)
    # Remove overlapping contours
    contours = remove_overlaps(cnts)
    # Sort by size, keep only the first NR_CHARACTERS
    contours = sorted(contours, key=lambda x: x['w'] * x['h'],
                    reverse=True)[:NR_CHARACTERS]
    # Sort from left to right
    contours = sorted(contours, key=lambda x: x['x'], reverse=False)
    return contours
```

```python
def extract_contour(image, contour, desired_width, threshold=1.7):
    mask = np.ones((image.shape[0], image.shape[1]), dtype="uint8") * 0
    cv2.drawContours(mask, [contour], -1, (255, 255, 255), -1)
    result = cv2.bitwise_and(image, mask)
    mask = result > 0
    result = result[np.ix_(mask.any(1), mask.any(0))]

    if result.shape[1] > desired_width * threshold:
        # This contour is wider than expected, split it
        amount = ceil(result.shape[1] / desired_width)
        each_width = floor(result.shape[1] / amount)
        # Note: indexing based on im[y1:y2, x1:x2]
        results = [result[0:(result.shape[0] - 1),
                          (i * each_width):((i + 1) * each_width - 1)] \
                   for i in range(amount)]
        return results
    return [result]

def get_letters(image, contours):
    desired_size = (contours[-1]['x'] + contours[-1]['w'] - contours[0]['x']) \
                   / NR_CHARACTERS
    masks = [m for l in [extract_contour(image, contour['c'], desired_size) \
             for contour in contours] for m in l]
    return masks
```

With this, we're finally ready to write our cutting script ("cut.py")

```python
from os import makedirs
import os.path
from glob import glob
from functions import *
from constants import *

image_files = glob(os.path.join(CAPTCHA_FOLDER, '*.png'))
```

```
for image_file in image_files:
    print('Now doing file:', image_file)
    answer = os.path.basename(image_file).split('_')[0]
    image = cv2.imread(image_file)
    processed = process_image(image)
    contours = get_contours(processed)
    if not len(contours):
        print('[!] Could not extract contours')
        continue
    letters = get_letters(processed, contours)
    if len(letters) != NR_CHARACTERS:
        print('[!] Could not extract desired amount of characters')
        continue
    if any([l.shape[0] < 10 or l.shape[1] < 10 for l in letters]):
        print('[!] Some of the extracted characters are too small')
        continue
    for i, mask in enumerate(letters):
        letter = answer[i]
        outfile = '{}_{}.png'.format(answer, i)
        outpath = os.path.join(LETTERS_FOLDER, letter)
        if not os.path.exists(outpath):
            makedirs(outpath)
        print('[i] Saving', letter, 'as', outfile)
        cv2.imwrite(os.path.join(outpath, outfile), mask)
```

If you run this script, the "letters" directory should now contain a directory for each letter; see, for example, Figure 9-20. We're now ready to construct our deep learning model . We'll use a simple convolutional neural network architecture, using the "Keras" library.

```
pip install -U keras
```

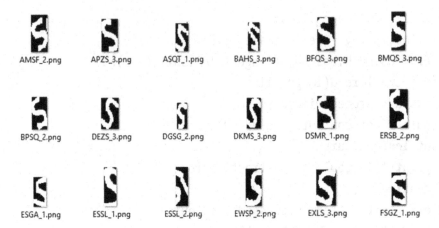

AMSF_2.png   APZS_3.png   ASQT_1.png   BAHS_3.png   BFQS_3.png   BMQS_3.png

BPSQ_2.png   DEZS_3.png   DGSG_2.png   DKMS_3.png   DSMR_1.png   ERSB_2.png

ESGA_1.png   ESSL_1.png   ESSL_2.png   EWSP_2.png   EXLS_3.png   FSGZ_1.png

Figure 9-20. *A collection of extracted "S" images*

For Keras to work, we also need to install a back end (the "engine" Keras will use, so to speak). You can use the rather limited "theano" library, Google's "Tensorflow," or Microsoft's "CNTK." We assume you're using Windows, so CNTK is the easiest option to go with. (If not, install the "theano" library using pip instead.) To install CNTK, navigate to *https://docs.microsoft.com/en-us/cognitive-toolkit/setup-windows-python?tabs=cntkpy231* and look for the URL corresponding with your Python version. If you have a compatible GPU in your computer, you can use the "GPU" option. If this doesn't work or you run into trouble, stick to the "CPU" option. Installation is then performed as such (using the GPU Python 3.6 version URL):

```
pip install -U https://cntk.ai/PythonWheel/GPU/cntk-2.3.1-cp36-cp36m-win_amd64.whl
```

Next, we need to create a Keras configuration file. Run a Python REPL and import Keras as follows:

```
>>> import keras
Using TensorFlow backend.
Traceback (most recent call last):
  File "<stdin>", line 1, in <module>
  File "\site-packages\keras\__init__.py", line 3, in <module>
    from . import utils
  File "\site-packages\keras\utils\__init__.py", line 6, in <module>
    from . import conv_utils
```

```
  File "\site-packages\keras\utils\conv_utils.py", line 3, in <module>
    from .. import backend as K
  File "\site-packages\keras\backend\__init__.py", line 83, in <module>
    from .tensorflow_backend import *
  File "\site-packages\keras\backend\tensorflow_backend.py", line 1, in
  <module>
    import tensorflow as tf
ModuleNotFoundError: No module named 'tensorflow'
```

Keras will complain about the fact that it can't find Tensorflow, its default back end. That's fine; simply exit the REPL. Next, navigate to "%USERPROFILE%\.keras" in Windows' file explorer. There should be a "keras.json" file there. Open this file using Notepad or another text editor, and replace the contents so that it reads as follows:

```
{
    "floatx": "float32",
    "epsilon": 1e-07,
    "backend": "cntk",
    "image_data_format": "channels_last"
}
```

Using Another Back End   In case you're using Tensorflow, just leave the "backend" value set to "tensorflow." If you're using theano, set the value to "theano." Note that in the latter case, you might also need to look for a ".theanorc.txt" file on your system and change its contents as well to get things to work on your system, especially the "device" entry that you should set to "cpu" in case theano has trouble finding your GPU.

Once you've made this change, try test-importing Keras once again into a fresh REPL session. You should now get the following:

```
>>> import keras
Using CNTK backend
Selected GPU[1] GeForce GTX 980M as the process wide default device.
```

Keras is now set up and is recognizing our GPU. If CNTK would complain, remember to try the CPU version instead, though keep in mind that training the model will take much longer in this case (and so will theano and Tensorflow in case you can only use CPU-based computing).

We can now create another Python script to train our model ("train.py"):

```python
import cv2
import pickle
from os import listdir
import os.path
import numpy as np
from glob import glob
from sklearn.preprocessing import LabelBinarizer
from sklearn.model_selection import train_test_split
from keras.models import Sequential
from keras.layers.convolutional import Conv2D, MaxPooling2D
from keras.layers.core import Flatten, Dense
from constants import *

data = []
labels = []
nr_labels = len(listdir(LETTERS_FOLDER))

# Convert each image to a data matrix
for label in listdir(LETTERS_FOLDER):
    for image_file in glob(os.path.join(LETTERS_FOLDER, label, '*.png')):
        image = cv2.imread(image_file)
        image = cv2.cvtColor(image, cv2.COLOR_BGR2GRAY)
        # Resize the image so all images have the same input shape
        image = cv2.resize(image, MODEL_SHAPE)
        # Expand dimensions to make Keras happy
        image = np.expand_dims(image, axis=2)
        data.append(image)
        labels.append(label)

# Normalize the data so every value lies between zero and one
data = np.array(data, dtype="float") / 255.0
labels = np.array(labels)
```

```
# Create a training-test split
(X_train, X_test, Y_train, Y_test) = train_test_split(data, labels,
                                  test_size=0.25, random_state=0)

# Binarize the labels
lb = LabelBinarizer().fit(Y_train)
Y_train = lb.transform(Y_train)
Y_test = lb.transform(Y_test)

# Save the binarization for later
with open(LABELS_FILE, "wb") as f:
    pickle.dump(lb, f)
# Construct the model architecture
model = Sequential()
model.add(Conv2D(20, (5, 5), padding="same",
        input_shape=(MODEL_SHAPE[0], MODEL_SHAPE[1], 1),
activation="relu"))
model.add(MaxPooling2D(pool_size=(2, 2), strides=(2, 2)))
model.add(Conv2D(50, (5, 5), padding="same", activation="relu"))
model.add(MaxPooling2D(pool_size=(2, 2), strides=(2, 2)))
model.add(Flatten())
model.add(Dense(500, activation="relu"))
model.add(Dense(nr_labels, activation="softmax"))
model.compile(loss="categorical_crossentropy", optimizer="adam",
metrics=["accuracy"])

# Train and save the model
model.fit(X_train, Y_train, validation_data=(X_test, Y_test),
        batch_size=32, epochs=10, verbose=1)
model.save(MODEL_FILE)
```

We're doing a number of things here. First, we loop through all images we have created, resize them, and store their pixel matrix as well as their answer. Next, we normalize the data so that each value lies between zero and one, which makes things a bit easier on the neural network. Next, since Keras can't work with "Q", "W",... labels directly, we need to binarize these: every label is converted to an output vertex with each index corresponding to one possible character, with its value set to one or zero, so that "Q" would become "[1, 0, 0, 0,...]," "W" would become "[0, 1, 0, 0,...]," and so on. We save this conversion as we'll also need it to perform the conversion back to characters again during application of the model. Next, we construct the neural architecture (which is relatively simple, in fact), and start training the model. If you run this script, you'll get an output as follows:

```
Using CNTK backend
Selected GPU[0] GeForce GTX 980M as the process wide default device.

Train on 1665 samples, validate on 555 samples
Epoch 1/10

C:\Users\Seppe\Anaconda3\lib\site-packages\cntk\core.py:361: UserWarning: ↵
your data is of type "float64", but your input variable (uid "Input4")  ↵
    expects "<class'numpy.float32'>". Please convert your data          ↵
    beforehand to speed up training.
  (sample.dtype, var.uid, str(var.dtype)))
  32/1665 [..............................] - ETA: 36s - loss: 3.0294 -
                                             acc: 0.0312
  64/1665 [>.............................] - ETA: 22s - loss: 5.1515 -
                                             acc: 0.0312

[...]

1600/1665 [=============================>..] - ETA: 0s - loss: 7.6135e-04 -
                                             acc: 1.0000
1632/1665 [=============================>.] - ETA: 0s - loss: 8.3265e-04 -
                                             acc: 1.0000
1664/1665 [=============================>.] - ETA: 0s - loss: 8.2343e-04 -
                                             acc: 1.0000
1665/1665 [==============================] - 3s 2ms/step - loss: 8.2306e-
                                             04 - acc:
      1.0000 - val_loss: 0.3644 - val_acc: 0.9207
```

We're getting a 92 percent accuracy on the validation set, not bad at all! The only thing that remains now is to show how we'd use this network to predict a CAPTCHA ("apply.py"):

```python
from keras.models import load_model
import pickle
import os.path
from glob import glob
from random import choice
from functions import *
from constants import *

with open(LABELS_FILE, "rb") as f:
    lb = pickle.load(f)

model = load_model(MODEL_FILE)

# We simply pick a random training image here to illustrate how predictions
work. In a real setup, you'd obviously plug this into your web scraping
# pipeline and pass a "live" captcha image
image_files = list(glob(os.path.join(CAPTCHA_FOLDER, '*.png')))
image_file = choice(image_files)

print('Testing:', image_file)

image = cv2.imread(image_file)
image = process_image(image)
contours = get_contours(image)
letters = get_letters(image, contours)

for letter in letters:
    letter = cv2.resize(letter, MODEL_SHAPE)
    letter = np.expand_dims(letter, axis=2)
    letter = np.expand_dims(letter, axis=0)
    prediction = model.predict(letter)
    predicted = lb.inverse_transform(prediction)[0]
    print(predicted)
```

If you run this script, you should see something like the following:

```
Using CNTK backend
Selected GPU[0] GeForce GTX 980M as the process wide default device.

Testing: generated_images\NHXS_322.png
N
H
X
S
```

As you can see, the network correctly predicts the sequence of characters in the CAPTCHA. This concludes our brief tour of CAPTCHA cracking. As we've discussed before, keep in mind that several alternative approaches exist, such as training an OCR toolkit or using a service with "human crackers" at low cost. Also keep in mind that you might have to fine-tune both OpenCV and the Keras model in case you plan to apply this idea on other CAPTCHA's, and that the CAPTCHA generator we've used here is still relatively "easy." Most important, however, remains the fact that CAPTCHA's signpost a warning, basically explicitly stating that web scrapers are not welcome. Keep this intricacy in mind as well before you set off cracking CAPTCHA's left and right.

Even a Traditional Model Might Work   As we've seen, it's not that trivial to set up a deep learning pipeline. In case you're wondering whether a traditional predictive modeling technique such as random forests or support vector machines might also work (both of these are available in scikit-learn, for instance, and are much quicker to set up and train), the answer is that yes, in some cases, these might work, albeit at a heavy accuracy cost. Such traditional techniques have a hard time understanding the two-dimensional structure of images, which is exactly what a convolutional neural network aims to solve. This being said, we've set up pipelines using a random forest and about 100 manually labeled CAPTCHA images that obtained a low accuracy of about 10 percent, though enough to get the answer right after a handful of tries.

Index

A

Amazon reviews, 198
  browser's developer tools, 241–242
  dataset library, 245–248
  "json" module, 243, 245
  Learning Python by Mark Lutz, 241
  matplotlib library, 250
  ntlk, 249
  POST requests, 241, 243
  Python shell, 249
  sentiment plots, rating level, 251
  text editor, 243
  vaderSentiment library, 248–249
Application Programming Interface
    (APIs), 4–5, 99, 180, 193, 197, 201
Associated Press (AP) vs. Meltwater, 182

B

Barclays' mortgage
    simulator, 214–215
Beautiful Soup, 67, 187–188, 190, 193
  attrs argument, 64
  code snippets, 71
  Comment objects, 76
  complex selectors, 75
  CSS selectors, 75
  documentation, 77
  elements, 74
  episode tables, 70

find and find_all, 65–66, 72
Game of Thrones, 68, 75
HTML pages, 61
HTML tree, 73–74
installing, 61
**keywords, 65
limit argument, 64
lists, and regular
    expressions, 73
methods, 63
name argument, 64
NavigableString objects, 76
parsers, 62–63
recursive argument, 64
string argument, 62, 64
tree-based representation, 63
warning, 62
Board members, 199, 278–281
Books to scrape, 197, 206–208

C

CacheControl, 188
Caching, 188–189
Cascading Style Sheets (CSS)
  Chrome's Developer Tools, 60
  formatting, 57
  HTML attributes, 56–57
  selectors, 58–61
  style declarations, 57–58
  Wikipedia page, 60

299

Get the eBook for only $5!

Why limit yourself?

With most of our titles available in both PDF and ePUB format, you can access your content wherever and however you wish—on your PC, phone, tablet, or reader.

Since you've purchased this print book, we are happy to offer you the eBook for just $5.

To learn more, go to http://www.apress.com/companion or contact support@apress.com.

Apress®

Printed in the United States
By Bookmasters